CW00607327

Building the ecological city

Building the ecological city

Rodney R White

CRC Press
Boca Raton Boston New York Washington, DC

WOODHEAD PUBLISHING LIMITED
Cambridge England

Published by Woodhead Publishing Limited, Abington Hall, Abington
Cambridge CB1 6AH, England
www.woodhead-publishing.com

Published in North America by CRC Press LLC, 2000 Corporate Blvd, NW
Boca Raton FL 33431, USA

First published 2002, Woodhead Publishing Ltd and CRC Press LLC
© 2002, Woodhead Publishing Ltd
The author has asserted his moral rights.

British Library Cataloguing in Publication Data
A catalogue record for this book is available from the British Library.

Library of Congress Cataloging in Publication Data
A catalog record for this book is available from the Library of Congress.

Woodhead Publishing ISBN 1 85573 531 8
CRC Press ISBN 0-8493-1379-1
CRC Press order number: WP1379

Cover design by The ColourStudio
Typeset by SNP Best-set Typesetter Ltd, Hong Kong
Printed by TJ International, Cornwall, England

Contents

Part IV Health: restoring urban ecosystem health

Preface

The current problem with the human system of resource use and residual disposal is that it is wildly out of equilibrium. Competition among humans is such that we slaughter thousands of our own species annually on the roads in our haste to get somewhere faster; we let millions of babies die every year for want of clean water and a modest diet; we deliberately kill more millions of people in war, often to secure access to dwindling resources. In the last fifty years we have become more aware that our pathological drive for 'more' has poisoned the land, the water and the air. We are changing the composition of the atmosphere and we are reducing the diversity of other species on which we depend.

The city is an opportunity to introduce some harmony into our place in the biosphere by requiring less space per person, by using resources more carefully, and by disposing of residuals more thoughtfully. So far we have not availed ourselves of this opportunity. On the contrary, the modern city relies upon a technology that is outstandingly profligate in its use of resources and production of unwanted residuals. If the modern city is a monument to anything, it is a monument to man's ecological inefficiency.

Behind the glittering façade of steel and concrete monuments to architectural ego, we have a human habitat in deep crisis – from the richest to the poorest urban system. At the family level the symbol of success is the automobile, although it clogs the streets, pollutes the air, and encourages members of the richest societies to indulge in 'road rage' – a primitive relapse into violence brought on by the frustration of traffic congestion.

The city evolved in ancient times as an opportunity for civilised discourse, the bedrock of democracy. Yet in the modern city we see a growing divergence between rich and poor. Economies continue to grow, more resources are used, yet the quantity of wasted humans – the destitute – also increases. Other visible signs of system failure in the modern city are rusting heaps of discarded automobiles and poisoned air. Some cities lie within a short distance of dramatic mountain ranges, but rarely can you see them from the city because the air we breathe is so foul. We may have banished the 'pea-souper' fog, but we have replaced it with a noxious mix of VOCs, SOx and NOx, which we extrude every day from our vehicles and power plants, and which we inhale as we respire. Our food and water are similarly transformed.

The irony is that it need not be like this. There is nothing inherently hopeless or wasteful about living at high densities. True, there are dangers if we are careless – but there are also dangers for the careless hunter-gatherer. On the contrary, high density is the *only* way for most of us to live, even at the current size of the human population. The trick is to learn to live smarter. We have to know when we are poisoning ourselves, and when we are taking vital resources from our desperate neighbours and from future generations.

Our problem lies in the fact that as our society becomes technologically more sophisticated it also becomes biologically more ignorant. We no longer know what we eat or drink, or where our wastes are taken. We do not know that there are mountains around the city because we seldom see them any more. We do not know that water runs under the city because it runs underground in sewers.

Building the Ecological City is an attempt to understand how we might begin to restore ecosystem vitality to modern urban systems. As such it might be characterised as a physical approach to our dilemma. Yet it would be a serious mistake to view the problem in such superficial terms. Our physical situation is a direct result of political and cultural choices – individual and collective. True, these might be taken by ourselves in ignorance, or by others against our will. Certainly the child living in the street did not choose to be without family or shelter, any more than the wealthy motorist chooses to spend three hours every working day in slow-moving traffic. Nonetheless, choices were – and are – made to *select* exactly those outcomes.

This book uses Abel Wolman's concept of 'urban metabolism' to explore how the modern urban ecosystem functions. It attempts to understand why the system malfunctions to produce pathologies such as urban blight and traffic congestion. Finally, it sketches some of the conditions required to restore the urban ecosystem to health – health to the people, and to the planet on which they depend.

Like its predecessor, *Urban Environmental Management* (Wiley 1994), this book owes a debt to many people in the field, many of whom are acknowledged in the text. Most of it was written while on sabbatical at the Environmental Change Institute of the University of Oxford, where I enjoyed the stimulating company of faculty, students and old friends, especially John and Brenda Boardman who included me in their field trips and many other interesting and enjoyable activities. Several of the publications of the Environmental Change Institute are cited in the text. It was an ideal circumstance in which to write a book. I was doubly fortunate to meet Iain Stevenson, my editor on the previous book, who was starting a new list for Woodhead Publishing. What was just an idea swiftly became a project.

I have since returned to the University of Toronto where I have continued to benefit from the encouragement and support of my friends and colleagues. I hope I have made all proper acknowledgements in the text. I have cited the work of several of my own urban students including Chris DeSousa, Eric Krause, T J Kulkarni, Pamela Robinson and Monica Tang as well as that of many of the students who have taken my graduate course, Cities as Ecosystems. I have profited on a daily basis from the input of many colleagues at the University and from the Adaptation and Impacts Research Group of Environment Canada with whom we have had a working partnership for over five years. I have particularly benefited from the advice and encouragement of Dave Etkin who introduced me to the world of the property and casualty insurance industry from the perspective of a meteorologist. I also learned a lot from setting out with Peter Timmerman to review the literature on coastal cities and climate change only to find that it did not seem to exist.

On a personal note I must thank Mona El-Haddad at IES who kept the world running around me so that I could finish the book. Without her help I would not now be at the finishing post. Most of all I would like to thank my wife, Sue, our daughters, Kathryn and Alyson, and Mum, for their encouragement. Sue also provided several of the photographs. All families must at some time resent the impact of book-writing on their lifes. However, they remained cheerful and very supportive, and for that I am very grateful.

I hope the outcome is some recompense for the effort that went into it from so many sources.

Every effort has been made to trace and acknowledge ownership of copyright. The publishers will be glad to hear from any copyright holders whom it has not been possible to contact.

Rodney White

Acronyms

ABI	Association of British Insurers
BP	British Petroleum
CAT	Centre for Alternative Technology
CCP	Cities for Climate Protection (ICLEI)
CDM	Clean Development Mechanism
CHEJ	Center for Health, Environment and Justice
COP	Conference of the Parties (FCCC)
DETR	Department of the Environment, Transport and the Regions (UK)
EPA	Environmental Protection Agency (USA)
FCCC	Framework Convention on Climate Change (UN)
ICLEI	International Council for Local Environmental Initiatives
IES	Institute for Environmental Studies, University of Toronto
IPCC	Intergovernmental Panel on Climate Change
ISO	Insurance Services Office (USA)
IULA	International Union of Local Authorities
JI	Joint Implementation
NOAA	National Oceanic and Atmospheric Administration (USA)
OECD	Organisation for Economic Cooperation and Development
OGWDW	Office of Groundwater and Drinking Water (USA-EPA)
OISE	Ontario Institute for Studies in Education, University of Toronto
TNR	Toronto–Niagara Region
UNEP	United Nations Environment Programme
UNFCCC	United Nations Framework Convention on Climate Change
USAID	United States Agency for International Development
USDA	United States Department of Agriculture

Other abbreviations

C	carbon
CFCs	chlorofluorocarbons
CH_4	methane
CO_2	carbon dioxide
DALY	disability-adjusted life years
DNA	deoxyribonucleic acid
GHG	greenhouse gas
GWP	global warming potential
HFCs	hydrofluorocarbons
HVAC	heating, ventilation and air conditioning
IQ	intelligence quotient
ISWM	integrated solid waste management
kg/m^3	kilograms per cubic metre
km	kilometre
kWh	kilowatt hour
mm	millimetre
MSW	municipal solid waste
MW	megawatt
NGO	non-governmental organisation
NIMBY	not-in-my-backyard syndrome
N_2O	nitrous oxide
NOx	nitric oxide and nitrogen dioxide
O_3	ozone
p	pence
PAHs	polycyclic aromatic hydrocarbons
PCBs	polychlorinated biphenyls

PCDDs	polychlorodibenzodioxins
PCDFs	polychlorodibenzofurans
PFCs	perfluorocarbons
PM_{10}	particulate matter, 10 microns or smaller
ppb	parts per billion
PV	photovoltaics
R–A–R	resources–activities–residuals
SCADA	Supervisory Control and Data Acquisition systems, for water supply
SF_6	sulphur hexafluoride
SOx	sulphur oxides
TB	tuberculosis
$\mu g/m^3$	micrograms per cubic metre
UPS	uninterruptible power supply
UV-B	ultraviolet radiation, within the range of 280 to 320 nanometres
VAT	value added tax (UK purchase tax)
VOCs	volatile organic chemicals

Note: Measurements throughout the book are metric unless noted otherwise. The dollar sign signifies US dollars unless noted otherwise.

Part I
Introduction

1

Cities for the new millennium

We are convinced that sustainable human life on this globe cannot be achieved without sustainable local communities. Local government is close to where environmental problems are perceived and closest to the citizens and shares responsibility with governments at all levels for the well-being of humankind and nature. Therefore, cities and towns are key players in the process of changing lifestyles, production, consumption and spatial patterns. (The Aalborg Charter 1994)

It is possible to design living technologies that have the same capabilities as natural systems do – self-design, self-repair, reproduction and self-organisation in relation to changes – functions that now take technological society inordinate amounts of chemicals, materials, and energy, often with harmful environmental consequences. (Todd 1996: 41)

1.1 Cities as a metaphor for Western society

In this book 'the city' is a metaphor for modern industrial life, which is based on an unsustainable level of use of natural systems. The term 'ecological city' may appear strange to some readers, perhaps suggestive of a gingerbread house, something edible, ephemeral, even quirky. Yet an ecological city is the most durable kind of settlement that humans could build, if it is defined as 'a city that provides an acceptable standard of living for its human occupants without depleting the ecosystems and bio-geochemical cycles on which it depends'. That sounds like an entirely reasonable goal for human endeavour, something that you might hope we have striven for all along. Indeed one could accept at once that an 'ecological city' is much more than a metaphor, because it defines an important human goal. After all, there is little point in building cities that *do* deplete the physical environment on which we depend.

So why have we done exactly that? The short answer is that we did it unknowingly, just as we continue, even now, to stride confidently into an unknown future.

1.1.1 Four assumptions and a proposition

In this book it is assumed that climate change is a real threat. Even while the book was in the process of being written, opinion continued to strengthen behind the belief that the implications of climate change will be significant and that we are already experiencing the first stages of a long term shift towards a warmer world. It is also assumed that the global water crisis will grow in magnitude. Climate change occurs because the global carbon cycle has been disturbed by human action. Similarly the water crisis can be seen as a disturbance of the global hydrological cycle. A third key assumption is that national governments, and the international bodies to which they belong, will find it very difficult to respond to the challenge of climate change. Fourth, a balancing assumption is that urban communities (their governments and citizens) have both the means and the motivation to make a significant contribution to our response to climate change and to other environmental problems. They have the technical means lying to hand – there is ample evidence of that. They have the motivation because cities are very vulnerable to climate change and to the water crisis.

There are several approaches to defending the fourth assumption. Take, for example, a statement from the Charter of European Cities and Towns Towards Sustainability, also known as the Aalborg Charter:

> We are convinced that the city or town is both the largest unit capable of initially addressing the many urban architectural, social, economic, political, natural resource and environmental imbalances damaging our modern world and the smallest scale at which problems can be meaningfully resolved in an integrated, holistic and sustainable fashion. As each city is different, we have to find our individual ways towards sustainability. (Aalborg Charter, paragraph 1.3 – see Appendix 1 of this book for the complete text)

The arguments in this book support an encouraging proposition to the effect that cities that become environmentally sensitive, or eco-friendly, will enjoy a competitive advantage that will increase over time as the global environmental struggle intensifies. For example, society cannot long delay the imposition of significant taxes on energy derived from fossil fuels – so-called carbon taxes. Therefore any city that offers an energy-efficient environment for business and residence will be offering tangible, bottom-line returns. Similar advantages will accrue to the efficient use of water and other materials.

1.1.2 A new era in urban risk analysis

As the world continues to urbanise people become more and more concentrated spatially. There are many commercial and social advantages to the accessibility that this concentration offers. There are also several risks inherent in living at high density. There is greater exposure to diseases, especially those that travel through contagion (see section 1.2) As cities become larger they also present larger targets to extreme weather events and earthquakes, greatly increasing the potential for disaster. In the context of exposure to hurricanes it is interesting to note that the current population of Miami, Florida, exceeds the total population in 1930 of all 190 coastal counties from Texas to Virginia. Cities are increasingly dependent on electricity and modern services for their everyday survival.

In the context of energy use and climate change cities are both major perpetrators of the problem and potentially major victims of the consequences. This provides them with a motivation both to adapt to the environmental changes to which we are already committed and to reduce further imbalances as rapidly as possible.

1.1.3 An organic analogy

The argument of this book depends on the use of an organic analogy. Cities are part of the natural environment. However artificial they have become, their inhabitants ultimately depend on the provision of clean air, clean water and healthy food. Cities are both systems of people, and ecosystems operating with the biosphere. Phrases such as 'cities as ecosystems' and the 'ecological city' reflect this assumption. There are two advantages to be gained from the use of this analogy. First, it makes it easier to think in terms of everything being linked to everything else, as in natural systems. At a certain level, cities act as a single system, as we see when they are stressed by extreme weather or other natural disasters. Second, the analogy helps us to think in terms of root causes rather than treating symptoms separately. There is little point in having one department responsible for cleaning the air and another for cleaning the water, if the pollutants are coming from a common source. Such 'end-of-pipe' solutions tend to be expensive and ineffectual.

The root cause of our imbalance with the natural ecosystems and biogeochemical cycles on which all life depends is the decision to expand our use, or throughput, of resources with no thought of the consequences. This is the way our modern, Western society has evolved, as we have gradually become more and more detached from the natural world.

1.1.4 Our divorce from biological reality

In the richer countries of the world, the daily existence of urban dwellers is separated from the natural ecosystems on which all life depends.

Water comes from a tap, milk from a bottle. Wastes go down the drain, up in the air, or into a dustbin. They go away to places that the producers of the wastes have never seen or thought about. Recently, some of these wastes have come back to haunt us. We have to relearn how natural ecosystems function, as Howard Odum proposed, so that we can remould our societies to mimic ecosystems and reuse our wastes (Odum 1971, 1983). This will require nothing less than a paradigm shift in urban planning and management.

It is 9000 years since the earliest known urban settlement thrived at Catalhöyük in present-day Turkey. For most of those 9000 years towns and cities served as ancillary settlements for a small minority of the population while the vast majority worked on the land. Cities provided a focus for administrative, military, religious and trading functions and they were dependent on the production of a surplus of agricultural produce most of which came from the surrounding land. Treatment of wastewater was non-existent and solid waste matter accumulated *in situ*; not surprisingly, death rates were high. Urban populations could be maintained only by a continual influx of people from the countryside.

While it is true that great empires like China and Rome supported long distance trade, even for staple cereals, the links between city and countryside, people and biosphere (the habitable parts of the planet) were very obvious. Urban people usually walked to their destinations. Those who were transported used human or animal power. They could see the surrounding countryside and had regular contact with it; many had been born there. In a sense these settlements functioned perfectly well according to ecological principles. The accumulation of wastes and pathogens kept human population growth rates down, while the productive capacity of the local agricultural region remained a key factor for urban continuity.

It is only in the last 250 years that this basic balance between the human population and the biosphere has been disturbed. Half the world's people – including almost all the rich – now live in urban settlements. Of these urban people some have never *seen* the countryside, few in the rich countries have ever lived there except as commuters or holiday-makers, and most are unaware of their continued dependence upon it. The philosopher Ingrid Stefanovic expressed the situation as follows:

> By virtue of the perceptual lack of contiguity between nature and urban form, it is easy for me to slip into a mind-frame that separates city from countryside as if indeed they were distinct entities – independent geographical containers of either urban or natural environment experiences. (2000: 45)

Despite this mental dislocation from our dependence on the natural functioning of the biosphere, we humans have continued to be a part of it in a physiological and biochemical sense, as sketched in Fig. 1.1. We are there in the top right hand corner, a population of mammals, living in com-

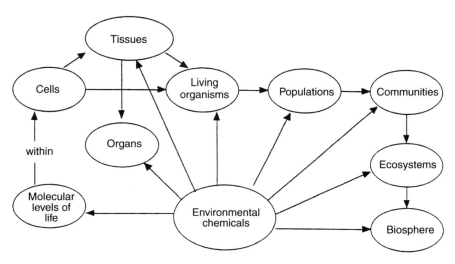

1.1 Levels of biological organisation
Source: after Connell et al. 1999.

munities, such as cities. Further up the biological scale we have transformed ecosystems and have even changed the composition of the atmosphere, a major component of the biosphere.

We have developed an awesome capacity to chop down trees, scoop up the world's fish, and unleash enormous quantities of energy by burning fossil fuels. We have also reproduced ourselves on a scale that must be the envy of lesser pests. However, one thing that we have not learned to do is reflect on the path that we have taken. In our haste to seize the goods that we want we have rarely thought about the impact of the wastes that we do *not* want. Yet, as our use of resources has increased, so has our production of wastes of all kinds – solid, gaseous and liquid. These wastes are sometimes referred to as 'non-product outputs', or entities for which there is no market (Bower 1977). A technical word for wastes is 'residuals' – materials left over from production and consumption. Our challenge is to produce fewer of these residuals and create markets for those that we do produce. In the meantime, residuals continue to exist *somewhere*; they take up space in various places that we refer to as 'sinks'. Sinks may be as small as a landfill or a mine tailings pond, or they may be as large as an ocean or the global atmosphere. Some of the material lying in sinks is inert and may be ignored, while other materials continue to play an active role in the environment, such as contaminants in the food chain, or greenhouse gases in the atmosphere.

All of this could have been predicted if we had remembered the obvious fact that we are still a part of this biosphere that we are transforming. We are just another 'producer–consumer ecosystem' as represented by Howard Odum in Fig. 1.2. It should be added that 'waste' is not a natural concept,

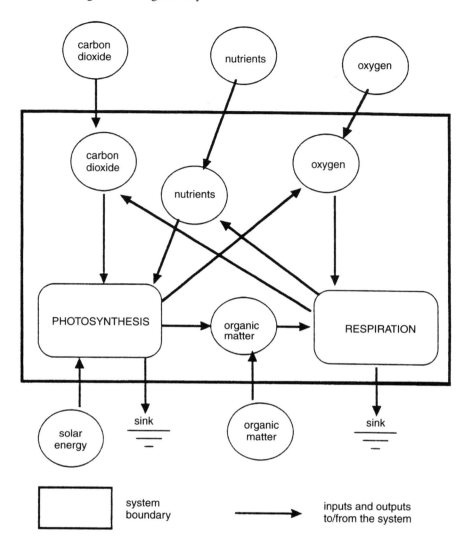

Arrows linking photosynthesis (of terrestrial plants, for example) and
respiration (of bacteria, for example) constitute a 'feedback'. Unusable
'residuals' from these processes are deposited in 'sinks' outside the system,
as defined by the boundaries.

1.2 A typical producer–consumer ecosystem
Source: after Odum 1983.

because, in natural systems, the outputs from one activity become the inputs
for other activities. As Connell *et al.* note regarding wastes in cities of the
ancient world:

It should be noted that the vast bulk of these wastes were natural
substances based upon products of the environment. We recognise

today that disposal of these wastes resulted in physicochemical and microbial-induced degradation, and that the wastes generated by relatively small community sizes could be readily handled by the receiving environments. (Connell *et al.* 1999: 3)

Thus the 'sinks' identified in Howard Odum's 'producer–consumer eco-system' are ecosystems in their own right which in turn produce their own residuals for export to other ecosystems.

1.1.5 Problems with residuals from human activities

Many of our current environmental problems derive from the fact that the sinks on which we depend cannot absorb all our wastes. The excess waste then disrupts the functioning of these sinks. Unfortunately, in the rich countries of the world, we do not have a large scale working model of an environmentally benign lifestyle, although there are plenty of examples of smaller communities that live by ecologically sound principles (Barton 2000; Gaia Trust and Findhorn Foundation 1996). We know what we need to do to move in that direction, but, so far, no rich society has taken the plunge. In the meantime, most inhabitants in the poorer countries know of only one lifestyle model that reduces infant mortality and improves quality of life. That model – whether developed under capitalism or communism – requires vast amounts of extra-somatic (or non-bodily) energy, most of which is derived from burning fossil fuels. It also requires huge inputs of other resources (principally land, minerals and water), while creating a related legacy of potentially troublesome waste materials.

Cities concentrate our wastes, thus producing huge landfills, polluted air and polluted water bodies. In the early stages of the industrial revolution the mess was accepted as an inevitable by-product of the new technologies. When the health problems of the poor (such as cholera and typhoid) were understood to affect the health of the rich, steps were taken to clear away the waste from the streets and bring in fresh water from streams and reservoirs in the countryside. Eventually drainage systems were laid to take sewage downstream. Thus was born the 'public health idea' of the 1840s, in Britain, for example (Ashton 1992). It was to be another hundred years before public concern about air quality brought regulation to North America and Western Europe to curb the grosser forms of pollution from particulates. It took a further forty years before regulations were introduced to take lead out of gasoline and reduce the quantity of sulphur dioxide emitted from smokestacks.

We are now facing the global challenge of human-induced climate change due to the accumulation of greenhouse gases in the atmosphere. To understand the magnitude of this kind of change we need to borrow concepts from systems theory as applied to the major bio-geochemical cycles such as the hydrological cycle, and the carbon, sulphur and nitrogen cycles. We have to step back and recognise that human society is as much a part of these cycles as is any other component of the natural world. We take in oxygen and exhale carbon dioxide, just as we take in minerals and extrude them too – some the same day, some only in death. We can analyse all of our activities in urban areas to see how they might affect a particular bio-geochemical cycle, or, in the case of a complex problem such as climate change, a set of cycles. This should help us to see that all the impacts of our activities are linked, just as the ecosystems they affect are themselves linked. At this moment, the cycles that we have most seriously unbalanced are the carbon cycle and the hydrological cycle.

The notion of cycles links back to the definition of residuals. Only humans have managed to create unusable wastes, and these are the ones that are upsetting the natural cycles on which we depend. We must now re-examine all our activities to make sure that our waste products are put back into beneficial use.

1.2 The urban environment and human health

The environmental debate is sometimes side-tracked by the objection that 'people are more important than the environment', as if the environment were something that existed apart from the human experience. This objection seems to be based on the assumption that environmental quality is a luxury that we can afford only when our basic (and some not-so-basic) needs have been met.

The arguments used in this book will be based exclusively on the self-interest of the human species, in the belief that environmental quality is what we need in order to maintain ourselves. Environmental quality is not a luxury, nor an option, but a necessity. Nowhere is this situation clearer than in the urban environment. Until modern times, urban mortality rates exceeded rural mortality rates. Partly this was due to the prevalence of pathogens that exist in human wastes, partly to disease vectors such as rodents, mosquitoes and flies. The density of urban settlements ensured that diseases that are spread by human contact enjoyed an ideal breeding ground, encouraged further by the even higher density of people living in crowded rooms.

We can trace a mortality transition from the early days of urban settlements when deaths due to environmental factors (especially inadequate water and sanitation) predominated (Meade *et al.* 1988; Rowland and Cooper 1983). In modern, Western cities with much lower mortality rates, a majority of deaths are attributed to cancer and heart disease. We have also introduced new threats through the gases, minerals and synthetic chemicals that have been emitted into the air, land and water, such as ozone and lead.

This book is about the interface between urban metabolism and human metabolism. Many of the same materials that cycle through the urban environment also cycle through the human body. Phrases such as the 'healthy city', the 'green city', and 'environmental health' try to capture this interplay between material cycles outside the body and the cycles inside the body. Since ancient times philosophers have compared the functioning of the microcosm of the human body to the macrocosm of the earth itself (Gould 1999). Yet, in modern times, we have separated our attention to human health and urban planning, except when we consider specific standards, such as ambient air quality, drinking water standards, or housing density. Only recently has it been proposed that the key to improving the physical quality of urban life can be found in understanding the interplay between all these standards as *one inter-linked system*.

Above all, we have to develop a more coherent view of the *balance* of materials required for the healthy functioning of ecosystems, including the city, as the most complex example of a human ecosystem. It has been recognised since ancient times that balance is the key to healthy living. We were reminded in the sixteenth century by the physician, Paracelsus, that: 'All substances are poisons. There is none which is not a poison. The right dose differentiates a poison from a remedy.' Yet our modern technological society pays scant heed to this wisdom. Tests are carried out on specific new products that are then licensed for use under certain conditions. But the global, long term implications of their manufacture and release may take decades to observe (Munn 1994; Munn and Wheaton 1997). Some of the most notorious products – such as DDT, CFCs and asbestos – are then recalled or subjected to further restrictions.

Environmental threats to human health include natural pathogens that proliferate in human and animal wastes (e.g. *E. coli* and *Cryptosporidium*) and toxicants manufactured by humans, such as petroleum, industrial materials (e.g. PCBs, CFCs), pharmaceuticals and biocidal agents, such as pesticides and herbicides. In addition to the natural pathogens and the manufactured materials, humans have also amplified the concentration of naturally occurring minerals (such as arsenic, cadmium, mercury and lead), salts, plant nutrients (including compounds of nitrogen and phosphorus), and gases like ground-level ozone, acidifying compounds, and greenhouse gases. Sources of potential impact on environmental health are summarised in Table 1.1.

The impact of toxicants has been observed, even at low concentrations, throughout the world, most visibly on plants and animals. Common effects are carcinogenesis (causing tumours, some benign, some malignant), mutagenesis (leading to heritable mutation through alteration of DNA and chromosomes), and teratogenesis (malformation of embryos), as well as biochemical, physiological and behavioural responses. Identifiable impacts on humans are usually confined to extreme events, such as the release of methyl mercury at Minamata in Japan. Many other cases have produced epidemiological evidence that is disputed by those held responsible for the suspect discharges.

Examples of toxicants routinely discharged into the environment are listed in Table 1.2. Each toxicant finds its own target area as illustrated in Table 1.3.

Table 1.1 Pollutants affecting environmental health

1. Natural pathogens encouraged by the increased availability and concentration of biological wastes from humans and domesticated animals.
2. Introduction of manufactured chemicals into the environment, which may be toxic even at low concentrations.
3. Human concentration of the bioavailability of naturally occurring substances:
   - minerals
   - salts
   - plant nutrients
   - gases.

Table 1.2 Sources and types of toxicants discharged into the urban environment

Source	Some chemical groups involved
1. Motor vehicle exhausts, electricity generation and industrial discharges to the atmosphere	1. Lead and other toxic metals, carbon monoxide, carbon dioxide, aromatic hydrocarbons, sulphur dioxide, hydrocarbons, PCDDs, PCDFs, PCBs*
2. Sewage	2. Aromatic hydrocarbons, hydrocarbons, chlorohydrocarbons, toxic metals, surfactants**
3. Stormwater runoff	3. Aromatic hydrocarbons, hydrocarbons, lead and other toxic metals
4. Industrial discharges to waterways	4. Acids, toxic metals, salts, hydrocarbons, PCDDs, PCDFs, PCBs
5. Urban and industrial discharges to the soil	5. Toxic metals, salts, hydrocarbons, PCDDs, PCDFs, PCBs

* PCBs, polychlorobiphenyls; PCDDs, polychlorodibenzodioxins; PCDFs, polychlorodibenzo-furans.
** Solvents used for mobilisation.
Source: Adapted from Connell *et al.* 1999: 21.

Table 1.3 Examples of the physiological targets of some common toxicants

Toxicant	Target areas
1. Lead	Bone, teeth, kidney; circulatory, reproductive and nervous systems
2. Mercury	Nervous tissue
3. DDT, PCBs	Fat, milk
4. Aflatoxin, vinyl chloride	Liver
5. Asbestos	Lungs

Source: Adapted from Connell *et al.* 1999: 55.

It is sometimes implied that the worst environmental problems have now been solved in the West, and that other countries will experience a similar success once they attain a Western standard of living. There are several reasons why this complacency is misplaced. First, the impact of many pollutants is long term, as has been experienced on numerous occasions. As human beings occupy the top of the food chain they may be the last to be affected. Other effects are difficult to detect, although widespread, and only come to light when long term studies are compared on an international basis. Some communicable diseases (such as TB) are reappearing in richer countries, as environmental refugees become more numerous. It is equally alarming to note that drinking water standards are being threatened around the world by pathogens like *E. coli* and *Cryptosporidium* against which water supplies in richer countries were supposed to be protected. Lastly, air pollution is now a global problem, with metals (such as mercury and cadmium) being carried from continent to continent. The most ubiquitous atmospheric problem is climate change induced by the greenhouse gases released by human activities, and this will have widespread effects on human health. The details of these impacts and the range of possible responses will appear in subsequent chapters.

1.3 The urban management challenge

1.3.1 The ecological city

The challenges outlined above suggest that we need an ecological approach to urban design and management. Planners, politicians and citizens should reconceptualise cities *as ecosystems*, in order to mesh urban activities with the natural world on which we still depend. In other words, the city should function in harmony with the natural world. We should discard the notion that we can export our residuals to other places. Instead we should accept the inherent circularity of physical processes and begin to integrate our residuals as a human responsibility to the rest of the biospheric web.

One of the earliest models, in modern times, for this kind of approach was provided by Abel Wolman who introduced the concept of urban metabolism that focused on the major physical inflows to the city and outflows from it (1965). Wolman wanted to draw attention to the fact that managers of American cities in the 1960s made no conscious attempt to manage these flows, even though it was within their ability to do so. As a result, they suffered from interruptions in the supply of resources, such as water, and from the accumulation of residuals, such as air pollutants (Fig. 1.3).

Similar concepts flowed from Wolman's observation, including 'residuals and environmental quality management' developed in the 1970s (Bower 1977), and this author's simplified concept of 'resources–activities–residuals' (Fig. 1.4). The first of these concepts made explicit the fact that

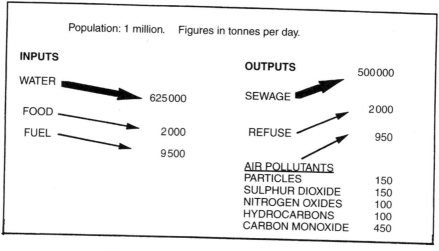

1.3 Wolman's hypothetical American city
Source: after Wolman 1965.

all economic activities produce unwanted residuals that must be managed effectively if environmental quality is to be maintained. The second concept is an attempt to develop a readily comprehensible notion of how these complex interactions can be managed. We need to be aware that we make choices all the time as to which resources we use, in pursuit of which activities, and which residuals we produce by making this choice. For example, when we drive a car we produce residuals that impair air quality and enhance the rate of climate change. When we insulate a house, or plant a shade tree, we reduce both these negative impacts. Examples are easy to find but it is difficult to relate individual behaviour to large scale environmental impacts.

However, there is a model that enables us to make the connection between the individual and the global. The 'ecological footprint' is based on an individual's pattern of consumption, aggregated into a single measure of the land required to support the individual's various activities, such as food and transport requirements, energy use, landfill requirements and so on. The first footprint study in Canada concluded that the average Canadian required 4.3 hectares of productive land to support his or her annual needs, thus prompting the observation that we would need more than three 'planet Earths' to support *even the current world population* at a level of consumption typically found in rich countries (Wackernagel and Rees 1996: 15). Footprint studies have been extended to international comparisons and to specific cities (Wackernagel *et al.* 1999; van Vuuren and Smeets 2000; Krause 1997; Onisto *et al.* 1998).

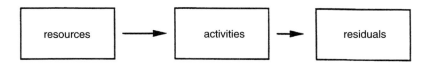

The original input – output concept was developed to measure the effect of the economy on itself, on the basis of fixed, linear technical coefficients. The analysis produced quantities of pollution, whose first order impacts were added up as costs and benefits.

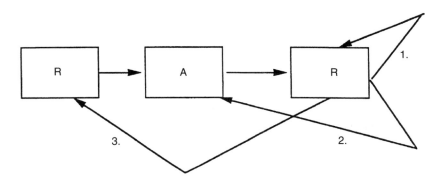

KEY
1. Interaction between residuals, e.g. volatile organic compounds and ozone.
2. Impacts of residuals on human activities, e.g. human health and sea level rise
3. Residuals attack resources, e.g. the effect of ultraviolet radiation on phytoplankton.

1.4 The resources–activities–residuals approach

Several similar themes are illustrated by such phrases as 'healthy cities', 'green cities', 'eco-cities', and 'eco-communities' (Ashton 1992; Roseland 1997, 1998; Smith *et al.* 1998; Barton 2000). All of these approaches emphasise the physical dependence of urban systems on the environmental services provided by the natural world. To this recognition they add social objectives such as human health, employment and quality of life. Finally they link these concepts to the broader search for 'sustainable development', being a form of development that can satisfy the needs of the present world population without compromising the satisfaction of the needs of future generations (Haughton and Hunter 1994; Stren *et al.* 1992).

In the course of reading this book the reader is encouraged to calculate his or her own ecological footprint using the City of Toronto 'ecological footprint calculator' (listed with the websites at the end of this chapter).

The footprint index, and the concept of resources–activities–residuals, can be used as guideposts through what sometimes is in danger of becoming a complex computational exercise. Their use should demonstrate opportunities to implement what might be termed a 'virtuous cycle, or hierarchy' of reduce, reuse and recycle. The application of the right technology is a part of this endeavour, but the 'right', or more appropriate, technology will require changes in human priorities and behaviour. Improved urban design and management should be based on a realistic assessment of transitional states of human adaptation to the challenges we face. We can see from the violent, chaotic and angry response to petrol price increases in Western Europe (in September 2000) that this will not be easy. Adaptation will have to proceed in an incremental fashion if it is to gain popular acceptance. It must also be accompanied by environmental education at all levels from elementary to adult.

Human–environment interactions are complex, but the key principle is terribly clear: humans – like all other species – must learn to live within the biological realities of the planet. We can run, we can shut our eyes, but we cannot hide.

1.3.2 Why the climate change problem changes everything

There has been growing concern since the 1950s that human production of additional greenhouse gases (such as carbon dioxide and methane), that maintain the planet's warmth, could disturb climatic regimes on a global scale (Firor 1990). This disturbance will not only raise surface temperatures, but will also raise the sea level and produce more extreme weather events. The reasons for concern reached a wider audience with the formation of the Intergovernmental Panel on Climate Change, in 1988, under the joint auspices of the World Meteorological Organisation and the United Nations Environmental Programme (Houghton 1994). Each successive report from the IPCC has provided additional confirmatory data on what climate change might mean for human society and for the other inhabitants of planet earth (Harvey 2000; Houghton *et al.* 1990, 1992, 1996). The growing pressure provided by this knowledge is an additional reason why we should adopt a more holistic approach to urban design and management.

Concern over climate change was one of the driving forces behind the United Nations Conference on Environment and Development held in Rio de Janeiro in 1992. Two major outputs from that conference, of relevance to the themes addressed in this book, are the Framework Convention on Climate Change and Agenda 21, the latter comprising 'a set of goals and programmes for environmental sustainability endorsed by 150 nations' (Gilbert *et al.* 1996). The expectation at Rio was that most local authorities throughout the world would be inspired to develop a 'Local Agenda 21' to improve quality of life at the local level (see http://www.iclei.org/rioplusten/index.html#survey). Reference to that goal will be made throughout this book because not only does climate change have important implications for the quality of urban life, but also urban life has important implications for the evolution of climate change. What climate change will eventually mean for human society, and the rest of the planet, depends on what human beings decide to do across a whole range of their activities, especially those related to energy use, and the production of food, fibre and materials. As the urban percentage of the population reaches 80% in the richer countries, and continues to rise in the poorer countries, the nature of urban life becomes an increasingly important factor in the equation. We can describe cities as being vulnerable to climate change, but it is equally true that the way they function affects the way events unfold.

Even the largest cities make only a small contribution to the total global emission of greenhouse gases and thus their inhabitants may not feel particularly effective in making a difference. However, every city could make an important contribution through the demonstration effect of any improvement in the current situation. It might be useful to think of each city as a cell in the global body, whereby the development of an effective antibody could strengthen the whole of the system. At this stage the future is wide open because it is most unclear how seriously individuals and governments will respond to the climate change challenge. Simulation models are producing a wide range of scenarios that might describe the future, operating from a global to a regional scale. We can establish a range of futures running from best to worst, and one somewhere in the middle.

We might describe the 'best case' scenario as one in which we make a viable transition to a fossil fuel-free world. Perhaps, through a combination of regulations and market incentives, sustainable development becomes the business norm. The global population is ageing and slowly declining, as global co-operation ensures the steady reduction of poverty.

Under the 'worst case' scenario we stay with 'business as usual', with steadily mounting greenhouse gas emissions leading to what some observers have called a 'runaway greenhouse' situation. In the runaway greenhouse, higher temperatures melt the permanently frozen ground in the tundra, releasing immense quantities of methane, greatly enhancing the rate of warming; higher temperatures melt the ice-caps and accelerate the rate of sea level rise. This would lead to a high level of unpredictability in every activity, in a world buffeted by extreme weather events and awash with environmental refugees.

Between these two extremes we can posit an intermediate case in which 'mankind muddles on'. We would see some reduction in greenhouse gas emissions, but the gains would be outweighed by increases elsewhere, especially in the transportation sector. For example, low emission vehicles might become the norm, but as their number is rising rapidly they still continue to increase their contribution to warming, as well as to traffic congestion and very poor air quality. Smog and high ozone concentrations become a global problem, while water shortages become common even in rich countries.

It should be emphasised that while climate change is a relatively new and potentially severe problem, the processes that have created it are indicative of a pattern of human activity that has been established for more than 200 years. Even if climate change were to recede as a threat, these underlying problems – such as pollution and water shortages – would remain. In the meantime, climate change is a tie that binds all the world's countries and their people together. *There is no solution to this problem without a fundamental agreement between rich and poor countries on how we should proceed.* For the first time in human history the victors can no longer seize the spoils and disregard the consequences.

1.3.3 Breaking the sectoral mould

There are many reasons why our industrial-age technology has evolved in a reactive, careless fashion, the main one being that our budget, design and management functions – in both the public and private domains – are organised primarily by separate sectors or activities, such as transportation, education, and waste management, with relatively weak liaison between them. Managers have long recognised this situation and efforts are constantly undertaken to provide the necessary integration, albeit with very limited success.

The sectoral approach to urban design and management was supported by the assumption that a problem that had been relocated from an area of concern was a problem that had been solved. If local supplies of water or power were deficient then they could be replaced by supplies from further afield. Wastes that could not be handled locally were sent to distant landfills; air pollutants were dispersed by tall chimneys. These 'solutions' for the city became problems for another locality – valleys flooded for hydro-electric power, rural landfill sites, and distant regions blighted by acid deposition.

At the same time densely packed households were decanted from the city centre to the outskirts of the city, while richer people spread out further at lower densities in order to enjoy the amenity value of a detached, or semi-detached, single family dwelling. Their amenity value was protected by zoning for segregated land use, with vast areas reserved for housing, with few services or employment opportunities nearby.

The further the suburbs spread out, and the lower residential densities fell, the less viable public transport became. The distant suburbs were automobile-dependent. This approach was enshrined in planning practice as zoning for mono-functional land use. It was, at best, an oversimplified solution to some of the pollution problems of the early phases of urban industrial development. It was profligate of energy use and generated far more problems than it solved (Richardson 1992). Suburbanisation encouraged a shift from public modes of transport, walking and cycling, to the private use of the automobile for commuting to work. The indicator that describes the relative importance of the different modes of transport is known as the modal split.

If we now wish to change the modal split and encourage a return to public transport, walking and cycling, we will have to rebuild higher densities in the urban core and the suburbs. Intensification of land use could be accompanied by mixed land use that would provide housing, employment and services in closer proximity. Variety – in this case, variety of land use and economic and social functions – is an essential characteristic of a healthy ecosystem.

We could go further by reversing the present dependence on distant regions to support the city, by bringing demand and supply back together in the city, thereby making cities more self-sufficient. We can see examples of this happening already with the reuse of demolition waste to build foundations for new buildings on a site, in-house wastewater treatment using the Living Machine[1], and the development of power from waste biomass and from photovoltaic arrays installed on residential and commercial buildings.

In order to generate support for these kinds of initiatives we will have to abandon the sectoral mould in which our bureaucratic systems – both public and private – have evolved. The sectoral mould encourages competition for budgetary allocations and rewards actions that merely relocate problems. We need an organisational approach that encourages co-operation and leads to holistic solutions.

1.3.4 Linkages and partnerships

Because government cannot anticipate future needs and resources it must create conditions that allow society, as a whole, the flexibility to adapt to new circumstances. In this new socio-political landscape, functional linkages between different facets of society must evolve such that information can be transmitted and action taken (Gilbert *et al.* 1996). Any effective form of urban management must make these linkages and partnerships a priority. Only a few examples will be mentioned here as this theme runs throughout the book.

One vital functional linkage is that which exists between urban systems and the surrounding countryside. In the recent past, the powerful dynamic of the urbanisation process has simply taken rural resources for granted, with little consideration for the rights of rural people or indeed for the long term availability of those resources. Expressways simply *had* to be built and land *had* to be appropriated for disposing of urban waste and for building

[1] Living Machine is a trademark of Ocean Arks International.

low density housing. Even the air-sheds over rural areas are now contaminated by health-threatening ozone. It is ironic that, in this age of cheap
electronic communication, dialogue between rural and urban interests has
never been weaker, mainly because the overwhelming majority of the
population of the rich countries is now urban, and is becoming increasingly
ignorant of rural life.

The trend towards privatisation, deregulation and decentralisation is
another powerful force that has not yet run its course. Like urbanisation it
is a mixed blessing. In theory, it should be more flexible and responsive to
people's needs than the former state monopolies that provided many services (such as water supply and treatment) in Westernised industrial countries, throughout the communist world, and in the ex-colonial world too.
However, these services are still regulated by centralised state institutions,
so the distinction between public and private control is less clear than the
rhetoric suggests. Indeed, de-regulation from public sector to private sector
requires *re*-regulation to ensure that public oversight is effective. It is an
important theme of this book that, whatever the allocation of power
between public and private institutions, the same physical parameters laid
down by nature still exist. Privatisation cannot create more water – it simply
intercepts the natural hydrological cycle in a slightly different way. The
over-pumping of an aquifer has exactly the same physical effect whether
the pumping agent is a state monopoly, private water utility, or an individual
farmer or factory.

As mentioned above, there are many small scale examples of 'eco-
communities' where people have organised themselves to make their
lifestyle more environmentally benign. The big question is whether this
process of self-organisation can be replicated on a larger scale to meet the
needs of major urban systems. Can technologies such as the Living
Machine, neighbourhood composting, and photovoltaic solar power be
scaled up to meet the demands of a million or more people in an urban
area? Do these technologies have space requirements that cannot be met
in a built-up area? Can we expect the social cohesion and goodwill that
underlies small eco-communities to operate among ordinary inhabitants of
large cities? Could such self-organised groups form an effective interface
with local authorities? Can information technology be employed to develop
an effective communication matrix to monitor the physical functions of a
city, and thereby develop a prototype for an 'intelligent city' (White 1994:
161–4)? Could such a system be designed to complement the natural, self-
organising technology identified by John Todd in the introductory quotation to this chapter?

Whatever the political and technological framework we employ, the need
for dialogue between all the players and agencies remains paramount. Non-
governmental organisations provide strong support for dialogue among all
parties, public and private, collective and individual. In some countries the
voice of environmental NGOs has come to play a major part in maintain-

ing dialogue between the governors and the governed, between the planner and the planned. One international NGO of particular relevance to this book is the International Council for Local Environmental Initiatives (ICLEI) which links local governments around the world on such projects as Local Agenda 21 and the Cities for Climate Protection Programme (*http://www.iclei.org*). ICLEI co-sponsored the fourth Local Government Leaders' Summit on Climate Change, a meeting held at Nagoya, Japan, in November 1997, parallel to the Third Meeting of the Conference of the Parties to the UN Framework Convention on Climate Change, which produced the Kyoto Protocol to implement measures to reduce greenhouse gas emissions. The local government leaders signed the Nagoya Declaration that pledged their support to the emission reduction campaign and urged their national leaders to take firm action (see Appendix 2). A key element of the Declaration was the involvement of poor countries, as well as rich, in this campaign. Such involvement was specifically excluded from the Kyoto Protocol, and this exclusion – predictably – has led to delays in its implementation.

1.3.5 The international context

A distinction between rich and poor countries has been made several times. The phraseology is clumsy because short words and phrases gloss over a very complex web of different societies. The everyday adjectives 'rich' and 'poor' allude to a fifty-year evolution of attempts to distinguish between the central and the marginal countries in the modern world economy. It is interesting to note that the Kyoto Protocol produced a self-identified list of rich (or, at least, richer) countries as signatories for greenhouse gas reduction pledges. They are listed in Annex 1 to the Protocol and have thus become known as the 'Annex 1 countries'. The poor, by the same mechanism, have become known as 'non-Annex 1 countries' – a term scarcely less attractive or imaginative than the phrase 'underdeveloped countries' used in the 1950s. Ungainly as the new terminology may be, at least it is self-imposed and it derives directly from the different responsibilities and capacities of rich and poor countries. ('Annex 1 countries' are listed in Appendix 3 at the end of the book.)

The importance of the distinction for this book is that the environmental impacts of rich and poor are quite different (White 1993). The rich use more resources and produce more wastes. For example, a typical rich country inhabitant emits about 3.5 tonnes of carbon per year, compared with less than one-tenth of a tonne from a poor country inhabitant. (Specific examples of average emission loads are provided in Chapter 3.) However, the poor are still growing in number, so that we do not know if the ultimate global population will be 10 billion or 15 billion, or whether the population will stabilise at all. The importance of the total population 'at stability' is that the most likely circumstance that will produce stability is that the poor have become rich. On the present 'business as usual' trajectory this implies that their level of consumption and wastage will be similar to the rest of the rich, whatever that may become. Calculations such as the ecological footprint indicate that this type of scenario – 10 to 15 billion people, 60% of them owning a motor vehicle – is not a feasible future. Thus the linkages between rich and poor countries are a crucial factor in any calculation. Figure 1.5 describes this predicament as 'you can't get there from here'. The top right hand box represents the 'business as usual' dependence on fossil fuels and a high throughput of water and other materials, typical of the rich countries today, and the very development path being followed by poorer countries. The second part of the diagram suggests that poor countries should not dwell in that wasteful box too long, but take a short-cut to the top left hand box, which represents sustainable development. Unfortunately, it is not yet clear that this 'short-cut' exists.

The relevance of relations between rich and poor countries to options for urban management and design is another theme that runs throughout the book. For example, one global tie that has not yet been recognised is the flow of 'environmental refugees'. These are people who have left their country of origin because they perceived that there was no future for them there, and that even a dangerous journey and an uncertain future were preferable to staying put. Whatever steps are taken by rich countries to curb, or direct, these refugees, their volume will continue to increase until better conditions become available in their countries of origin.

An assumption of this book is that we already have the means to house, feed and water the world's people and that these means must be put into effect if we are to avoid a downward environmental spiral. Innovation could arise at any point in the global settlement system, and such innovations might be widely applicable because urban systems have much in common, at least from an ecological point of view. Cities need to share information on the innovations that appear to be promising. Some organisations, like ICLEI, already exist to facilitate this kind of international urban exchange. Yet – in terms of political influence – cities are almost invisible on the international environmental stage. This is something that will have to change if the rate of successful innovation – technological *and* social – is to be accelerated.

1.3.6 Building the ecological city

Ecological cities – if they are to be built at all – will be built mainly by people operating at the local level, not by planners or governments. Yet leadership is needed from all levels of government to make this possible. New incentive structures need to be put in place through international agreement and national legislation. 'Perverse taxes' that encourage waste need to be replaced by taxes, market structures, and local council decisions that encourage people to take steps to reduce their eco-

Where countries of the North and South are today

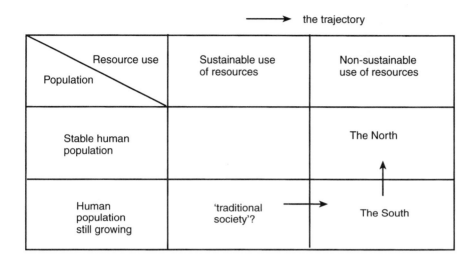

... and where we need to be

Resource use / Population	Sustainable use of resources	Non-sustainable use of resources
Stable human population	The North The South	
Human population still growing		

1.5 Where countries of the North and South are today . . . and where we need to be

logical footprint and develop lifestyles consistent with the resource limits of the planet.

Most of this change will be based on existing settlements, rather than new towns on greenfield sites. Thus the route to improvement lies through retrofitting the present building stock, factories and transportation systems.

Wherever possible, new structures should be built on land that has already been urbanised, so-called brownfield sites. If these sites are contaminated they will have to be remediated according to standards agreed by the builder, the government and the community. Technology that already exists, as proven 'best practice', could go a long way to making these changes a reality. What prevents more rapid improvement is the public's fear of having its lifestyle choices curtailed, matched by pervasive government paralysis, fearful of public outcry if new legislation or standards are imposed. A major purpose of this book is to demonstrate that such fears are mostly groundless. Any curtailment in individual choice will be small compared with the alarming uncertainties that we encourage by continued rapid population growth, depletion of critical resources like water, forests, soils, biodiversity and fisheries, and climate change.

It will become necessary to make regular use of indicators that reflect how well a city is performing over a representative range of activities, which together will represent the notion of urban ecosystem health. Indicators can be collected at all spatial scales from a national average to a city average, and down to surveys of the individual household. The most critical indicators are those that affect the major ecosystem flows, such as water use per head, carbon emissions per head, and quantity of solid waste generated per household. This book will identify the quantities that are involved in our daily resource use and the effect this use has on the functioning of the biosphere, following the old adage that 'you cannot manage what you cannot measure'.

1.4 How to use this book

1.4.1 Restoring urban systems to ecological balance

Clearly we have a problem. There are many well-understood indicators that tell us unequivocally that global patterns of resource use are not sustainable. Like the sorcerer's apprentice we do not seem to have any solution to hand. Our technological magic has a serious flaw. As Bill Vanderburg eloquently identified the problem:

> It is time to recognize that we are trapped in the labyrinth of technology. The metaphor seems appropriate since all too often we think we have successfully dealt with an issue, only to have it reappear somewhere else in a different form. The results are falling standards of living and the ongoing decline of all living systems. (2000: xi)

People playing many different roles in society are concerned about this situation, knowing that we do have the technical means to restore ecological balance to our cities if the political will can be established. Public responsibility for generating that political will rests with urban politicians and urban planners, and it is to them that this volume is directed.

However, they can only fulfil this role if they are supported by the rest of society, including the business community that operates largely in cities; environmental businesses, in particular, as they stand to gain from a more enlightened approach; the 'green community', including academics, environmental activists and green politicians; and mostly the 'ordinary citizen' who votes, pays taxes and lives in the urban environment that we have created.

This book is also addressed to this wider audience. It is designed and written in a style which should be accessible to people without technical training. It embraces the historical conditions that have brought us to this point, and it explores the future to which we are heading. It should be readable from cover to cover, or it may be used as reference only. Much is made of the value of a holistic approach to understanding urban ecosystems and it is hoped that some readers may absorb the emerging implications of this paradigm. However, for those immersed in particular, sectoral problems, such as water supply, or air quality, the chapters may be taken in a different order, as explained in the following section.

Some readers may want to use this book as a point of departure for more detailed analysis of a particular problem or approach, such as the ecological footprint or the Living Machine. To encourage such journeys each chapter concludes with suggestions for further reading, and many websites have been included, pointing readers to organisations that have developed solutions to some of the problematic environmental aspects of our urban systems. With the development of the web the distinctions between 'book–library–community–government' become blurred, as each takes on some aspects of the others.

No doubt different audiences will find different uses for this book. For the planners I hope it provides a useful manual. For the concerned citizen I hope it explains how we came to this particular point in our urban history and I hope it identifies some of the conditions necessary for the reduction of our more pressing problems, such as traffic congestion and climate change. However, we are a long way from actually building the ecological city. Thus, this book is more of a sketch than a blueprint.

1.4.2 Organisation of the book
This book is organised around two triads: one physical – land, air and water; the other conceptual – metabolism, pathology and health (see Table 1.4). The physical triad is deceptive in its simplicity, because it is the linkages between the environmental media that provide some of our greatest challenges. For example, it is the pollution of the atmosphere with greenhouse gases that produces global warming, yet the warming trend will almost certainly intensify the hydrological cycle. More rapid run-off, following heavier rainstorms, could in its turn release pathogens and contaminants more rapidly from the soil and from paved urban areas. However, this threefold

Table 1.4 Organisation of the book

Medium	Metabolism II	Pathology III	Health IV	V
Solid Waste & land	2	5	8	11 International issues
Gas Energy & air	3	6	9	12 The ecological city
Liquid Water	4	7	10	

division of the biosphere is something with which many people are very familiar and it should help to provide structure for the argument.

The conceptual triad of metabolism, pathology and health may be more problematic when applied to urban systems because these organic analogies are frowned upon in some circles. If cities are categorised as being simply the 'built environment' and hence inanimate, then the objection is valid, because the definition excludes cities from the realm of living things. However, the purpose of this book is to understand how cities interact with many living things. Indeed, it may be argued that it is more useful and accurate to envisage cities as systems of humans, organised through their individual and societal behaviour, rather than simply assemblages of bricks, mortar, cement and steel.

Metabolism is defined as 'the sum of chemical reactions that occur within living organisms' (Oxford University Press, 1992: 189), which in the urban analogy refers to physical flows into a city (mainly water, food and fuel), their transformation through production and consumption, and the extrusion of wastes (sewage, solid waste and pollutants to the air). This is the subject of Part II ('Metabolism: how urban ecosystems work'), in which Chapters 2, 3 and 4 focus, in turn, on land, air and water. The three chapters in Part III ('Pathology: what's gone wrong?') assess the problems that we have created through ignoring the physical parameters that govern our existence. 'Pathology' refers to malfunctioning cities, including physical blight, its impact on society, and the social conditions that produce it. Major examples of physical problems include contaminated urban land, poorly managed landfills, traffic congestion, inadequate air quality, the changing composition of the atmosphere (leading to climate change), water supply that is inadequate in terms of quantity and quality, and floods. It is emphasised throughout the book that these problems are simply the physical projections of social forces, and that individual behaviour and social relations are the keys to improvement. We cannot assume that the situation will be saved by a series of providential technological fixes.

Part IV is called 'Health', and is subtitled 'restoring urban ecosystem health'. It does not offer finished solutions to the problems presented in earlier chapters, but it does provide examples of innovations that seem to promise an improvement in the human condition without compromising the integrity of the biosphere for future generations. In Part V, two chapters draw together the main themes and conclusions. Chapter 11 returns to the international aspect of urban environmental issues, emphasising the need to share whatever innovations appear to work, given the fact that the condition of the rich is no longer isolated from the problems of the poor. The final chapter makes an assessment of the urgency and viability of possible human responses in the short, medium and long term.

Thus, 'metabolism' describes the physical flows that support a city; 'pathology' examines situations where those flows and systems malfunction; 'health' attempts to chart a corrective course. This may not produce cities that are sustainable until the end of time. It should, however, encourage us to design and manage ecological cities that function in harmony with the biosphere on which we ultimately depend.

1.5 Websites

1. Centre for Alternative Technology, UK: *http://www.foe.co.uk/CAT*

2. Centre for Neighborhood Technology: *http://www.cnt.org*

3. City of Toronto study of its ecological footprint:
   http://www.city.toronto.on.ca/energy/footprint.htm

4. City of Toronto ecological footprint calculator:
   http://www.city.toronto.on.ca/footprint/calc_fp.htm

5. Integrated urban environmental study of Greater Manchester:
   http://www.art.man.ac.uk/planning/cure

6. Institute for Sustainable Communities: *http://www.iscvt.org*

7. International Council for Local Environmental Initiatives:
   http://www.iclei.org/ – for Local Agenda 21 see:
   http://www.iclei.org/rioplusten/index.html#survey

8. Sustainable Communities Network:
   http://www.sustainable.org/index.html

1.6 Further reading

Aalborg Charter of European Cities and Towns Towards Sustainability (see Appendix 1).
Ashton, J, ed (1992). *Healthy Cities*. Milton Keynes, Open University Press.

Connell, D, Lam, P, Richardson, B and Wu, R (1999). *Introduction to Ecotoxicology.* Oxford, Blackwell Science.

Davis, M L and Cornwell, D A (1998). *Introduction to Environmental Engineering* (third edition). Boston, McGraw-Hill.

Douglas, I (1983). *The Urban Environment.* London, Edward Arnold.

Firor, J (1990). *The Changing Atmosphere: The Global Challenge.* New Haven, Yale University Press.

Gilbert, R, Stevenson, D, Girardet, H and Stren, R E (1996). *Making Cities Work: The Role of Local Authorities in the Urban Environment.* London, Earthscan.

Girardet, H (1992). *The Gaia Atlas of Cities: New Directions for Sustainable Urban Living.* New York, Anchor Books published by Doubleday.

Goudie, A (1986). *The Human Impact on the Natural Environment.* Cambridge, MIT Press.

Harvey, L D D (2000). *Climate and Global Environmental Change.* London, Prentice Hall.

Houghton, J (1994). *Global Warming: The Complete Briefing.* Oxford, Lion Publishing.

Jackson, A R W and Jackson, J M (1996). *Environmental Science: The Natural Environment and Human Impact.* London, Longman.

Roseland, M (1998). *Toward Sustainable Communities: Resources for Citizens and Their Governments.* Gabriola Island, BC, New Society Publishers.

Rowland, A J and Cooper, P (1983). *Environment and Health.* London, Edward Arnold.

White, R R (1994). *Urban Environmental Management: Environmental Change and Urban Design.* Chichester, Wiley.

White, R R and Whitney, J B (1992). 'Cities and the environment: an overview', in *Sustainable Cities: Urbanization and the Environment in International Perspective.* Eds R E Stren, R R White and J B Whitney. Boulder, Westview Press: 5–52.

Part II

Metabolism: how urban ecosystems work

2

It isn't waste until you waste it

The advantages of community living can be had . . . without paying the penalty of asphyxiation or starvation. The evolution of two-dimensional flat organisms or of open-mesh three-dimensional sponges illustrated two of the possible strategies. A more sophisticated solution to this problem came with the development of circulatory systems, with a network of channels and a means to sustain the flow of blood or plasma through them. (Lovelock 1991: 105)

2.1 Land use and urban metabolism

2.1.1 The density paradox

The problems referred to above by Lovelock relate to the evolution of the first multicellular organisms, yet the applicability to current issues in urban environmental management should be apparent. This is the essential density paradox of urban systems. Density is necessary for accessibility, but density implies a reduction in amenity and health if wastes are improperly handled. Traffic congestion wastes time and fuel, and impairs air quality. Eventually, untreated wastes accumulate in the ground and make the land unattractive for reuse, leading to abandonment and urban blight. These problems arise because the functioning of the urban system *as a whole* is ignored. What is needed is a holistic view of the physical functioning of the urban system.

Materials are brought into the city to supply industry and construction, food, clothing, and equipment for households. Although these materials are all 'used', they are not – from a physical perspective – consumed. They are merely transformed into other materials, many of which have no market value, and thus, for economic reasons, become wastes. The management of these wastes can become problematical in a number of ways. Some may be a direct threat to human health and thus classed as 'contaminants'; others reduce the amenity level of urban life through odours or unsightliness; all entail additional costs when they are moved out of the city.

The options for creating and handling these wastes are greatly influenced by the land use pattern that underlies the fabric of the urban environment. Most Western cities have developed into segregated land use systems with high value commercial uses in the centre, surrounded by transitional rings of industrial activities and lower valued residential areas, and, finally, expansive low density suburbs of single family dwellings. As this type of spatial structure evolved certain defaults became apparent. Journeys to work lengthened as suburban dwellings became distanced from jobs in the urban core. Zoning restrictions pushed industry out of the core to relocate near the suburbs on greenfield sites, leading to a dispersed pattern of jobs and increasing the typical journey to work. Public transport failed to take root in the low density suburbs so that families that could afford a second or third car eagerly embraced this opportunity for advanced consumerism. Some authorities taxed bus passes as a benefit while ignoring the provision of cheap or even free parking by employers; in Britain many companies still give their employees cars even when the job does not require it. Many people now drive into the urban core despite the availability of public transport. In some cities that once had dependable public transport, standards of comfort, convenience and safety have declined while prices have risen.

Patches of unused land began to appear in the downtown area and inner suburbs as industry moved out to cheaper greenfield sites. The unused land might become a temporary parking space but increasingly these 'brownfield sites' remained vacant through developers' fear of being held liable for any contaminants that might be found there. Banks began to regard these potentially contaminated sites as a liability rather than an asset that previously was valued as collateral for a loan to redevelop the site. These patches of urban blight began to cast a shadow on surrounding land, both residential and commercial. Meanwhile most of these cities continued to spread out – losing residential population density in the core, spreading the people ever more thinly into the surrounding countryside. There are exceptions to this picture but it has become typical of the Western city over the last fifty years. Why does it matter?

It has always been evident that there is a dynamic tension between the need to maintain residential population density at the core of an urban system and the problems caused by that density in terms of congestion, noise and potential for contamination. This is a problem inherent in the very nature of urbanisation and has been present since the beginning. Recently we have become aware of additional reasons for maintaining density in the urban system, specifically to support public transport as well as district heating and district cooling. At the same time we have become more aware of the need to preserve what remains of the surrounding green belt for recreation, flood control, food security, and air quality.

2.1.2 The material shift to the inorganic

Part of the current urban metabolic problem lies in the fact that the industrial revolution brought about a shift in our use of fuels and materials from the organic to the inorganic. We moved from a power system based on human and animal traction and fuelwood to one based on hydrocarbons and synthetic chemicals. For building materials we moved from wood to concrete and steel. Even the organic waste stream from humans may now be contaminated downstream with traces of elements like lead, cadmium, mercury, and phosphates and pesticides. Some materials that are now considered problematic – like lead and asbestos – were introduced because of their convenience. We now know that lead in paint, lead in water pipes, and lead in petroleum is a serious health risk, especially for children whose development may be permanently impaired by inhaling lead fumes from petrol or ingesting chips of leaded paint in their homes and schools. Likewise asbestos – once valued as a lightweight, malleable, fire-resistant building material – is known to cause deaths through lung disease thirty years after the fibres have been inhaled. Lead and asbestos are now being removed from buildings, at great expense, in every Western country.

Previously if a physical problem developed in a city then the rich moved away – upwind, upstream, or out of town. Traffic congestion was solved by building bigger highways. Polluting industry and garbage were moved away too. A problem left behind or relocated was considered to be a problem solved. Only recently have citizens and urban planners become critical of this approach. After all, it had served Western society very well through a period of unprecedented economic and demographic growth. The high costs of this casual, uncritical attitude are only just becoming widely apparent.

Poorer cities of the South that embarked on this same technological journey have been growing faster, demographically, with fewer public resources to spend on infrastructure or social services. For most people this means that urban systems provide polluted and inadequate water supply, hardly any water treatment, lethal air quality and widespread traffic congestion. There are exceptions, but they are few.

Now we can see that the modern approach to urban living is deeply flawed. In terms of energy and human health it is hugely inefficient and very expensive to maintain. When economic resources become strained the inadequacy of our modern cities as human life-support systems becomes painfully obvious. It is time to re-examine the physical basis on which these cities rest. The strange truth is that cities – as entire systems – are rarely examined from an efficiency point of view. Apart from three or four integrated studies the best we have are intermittent attempts to examine specific aspects of particular urban systems such as traffic flow, air quality, water supply and treatment, housing stock, and total energy use (Boyden *et al.* 1981; Ravetz 2000). On the other hand there are many examples of small scale, integrated ecological communities (Barton 2000).

The approach introduced in the previous chapter has the potential to provide an integrated overview of an urban system as a whole. The analogy of urban metabolism allows us to trace all the inflows into an urban system, to examine the efficiency with which these inflows support human activities, and to trace the 'chemical fate' of all the waste materials. Heat and noise are also important waste products. It is the combined magnitude of these inflows and wastes that determine the size of a citizen's ecological footprint. Understanding the nature of the flows is the first stage of a concerted attempt to reduce the size of that footprint to something that the earth can support.

2.1.3 Waste streams and waste management costs

The relative size of all urban metabolic flows – liquid, gaseous and solid – can be compared in Wolman's figures for the United States (Fig. 1.3) and Herbert Girardet's for London (Table 2.1). In terms of volume, every case shows the predominance of the water–sewage flow, followed in size by the gaseous conversion of oxygen to carbon dioxide, mostly in the combustion of fossil fuels. In terms of tonnage, solid wastes are by far the smallest component. However, the costs, per tonne, of managing the solid waste stream are by far the highest. The higher cost reflects the fact that, until recently, the extrusion of waste gases was completely ignored, being left to the agency of shifting air masses. Water treatment has been a significant cost for over 150 years in most Western cities, but even then it was the custom to discharge sewage, either after primary treatment, or even no treatment at all, to the nearest water body. However, nature has no force equivalent to mobile air masses or gravity driven hydrology, to remove solid wastes. In Western cities these wastes have been manually collected and mechanically buried for many decades. It is estimated in the United States that 70 to 85% of the cost of solid waste management is for collection, and of this collection cost 60 to 75% is for labour (Davis and Cornwell 1998: 636). This cost factor has meant that municipalities have been motivated to reduce the solid waste stream even before the environmental case had been made.

2.2 Sources and types of solid waste

2.2.1 Solid waste generation

In the post-war period in the West, urban inhabitants continued to use more and more materials to embellish a consumer lifestyle, with growing quantities of solid waste materials trucked out to be buried in landfills, despite

Table 2.1 The metabolism of Greater London (population 7 million) in millions of tonnes per year

	Inputs	Outputs
Solids		
Food	2.4	
Timber	1.2	
Paper	2.2	
Plastics	2.1	
Glass	0.36	
Cement	1.94	
Bricks, blocks, sand, tarmac	6.0	
Metals	1.2	
Industrial and demolition wastes		11.4
Household, civic and commercial wastes		3.9
Subtotal	17.4	15.3
Gases		
Oxygen	40	
CO_2		60
SO_2		0.4
NOx		0.28
Subtotal	40	60.68
Liquids		
Water	1002	
Wet, digested sewage sludge		7.5
Subtotal	1002	7.5

Source: Based on Girardet 1999: 33. Data compiled by the author for 1995 and 1996.

the deepening unwillingness of surrounding communities to host these projects. Piecemeal solutions were sought because there was no appreciation as to how the city functioned as a single metabolic entity – an entity that drew in materials, water and energy, and extruded waste elements to the ground, the water and the air.

The resources – activities – residuals approach provides a framework for understanding the nature of these flows (Table 2.2). All components of the flows provide opportunities to improve the efficiency of use, and reduce the quantity of waste that must either be secured inside the urban system or exported beyond its boundaries. Until recently all of these flows were examined and managed separately, usually distinguished by activity, rather than by material. Opportunities for internalising (i.e. within the urban administrative boundary) any of these flows were fortuitous or partial. For example, a landfill site might still be in operation within the urban boundary. Or local wastes might be incinerated within the boundary, even if the residual gases and ash were exported from it. Unofficially some waste producers, such as industries, simply buried their wastes on site, assuming that they would remain inert or, at least, undiscovered. Although statements are made from

Table 2.2 The 'resources–activities–residuals' approach applied to the solid waste stream in an urban system

Resources	Activities	Residuals
Food, including packaging	households	plastic, metals, paper, organic waste
Various	industry	various
Various	commerce and institutions	various
Building materials	construction industry	construction, renovation & demolition waste
Various	hospitals	special wastes
Solid fuel	power stations • thermal • nuclear • incinerators	flue ash nuclear waste flue ash

time to time concerning the intention of a community to manage 'its own' wastes, the details of how this might be done remain unclear, especially if it has no local landfill capacity left. Until recently many coastal cities were content to dispose of their solid waste at sea, New York City being one of the most notable of this number. Table 2.3 represents a typical urban solid waste stream.

The urban waste stream is a waste in several ways. First, we are wasting the materials themselves as well as the other inputs – including energy – used in the manufacturing process (see Table 2.4). Second, we are wasting more money, energy and materials in order to dispose of the waste. In addition, the disposal process can pose health hazards for humans, wildlife and vegetation. Disposal invariably involves a loss of amenity. Lastly, to the extent that we neglect the issue we are wasting an opportunity for people to learn about the material basis of their existence.

One difficult lesson that we have yet to learn is that as we become richer we create more wastes, because wastes are a linear function of the goods and services that we use. In fact you can predict the amount of solid waste an individual will generate as a function of his or her income, although it does vary also according to culture. In the United States the average figure is two kilograms per person per day, typically ranging from 1.5 to 2.5 kg according to various surveys (Davis and Cornwell 1998: 632). This is residential solid waste, the materials collected from the household. This does not include commercial and industrial wastes that are usually disposed of privately. As the commercial and industrial wastes are generated in order to provide individuals with goods and services, this should really be added to the 'per person' calculation, which would approximately double the total.

Table 2.3 Composition of a typical solid waste stream in a Western country

Component	Proportion by weight (%)
Paper	35
Garden waste	16
Food waste	15
Metals	10
Glass	10
Plastics, rubber, leather	7
Rags	2
Miscellaneous	5

Source: Jackson and Jackson 1996: 332.

Table 2.4 The potential savings from recycling (%)

	Aluminium	Glass	Paper	Steel
Water used	0	50	58	40
Energy used	90–97	4–32	23–74	47–74
Mining wastes	0	80	0	97
Polluting emissions to the atmosphere	95	20	74	85
Polluting discharges to watercourses	97	0	35	76

Sources: Data from UNEP, Environmental Data Report, 3rd edition, Blackwell, Oxford, 1991, presented by Jackson and Jackson 1996.

In addition to household, commercial and industrial wastes, there are certain types of waste that entail particular problems for disposal. For example, both hospital waste and radioactive waste require special handling because of the risks they pose to human health. Yet, until recently, these wastes were handled in what might be seen as a casual fashion. It was a common practice to incinerate hospital wastes *in situ*, despite the fact that most hospitals are located in urban areas, close to the centres of demand. As for radioactive waste, no country has an acceptable waste disposal system in operation, due to public mistrust of the solutions that have been offered, such as ocean disposal and sequestration in deep, stable geological formations like salt domes and granitic shields. Consequently, radioactive waste is most commonly stored in water tanks on the site of nuclear facilities, which, in some countries, are located close to urban areas.

Apart from these special cases of high health risks there is a general problem of the contamination of the waste stream by metals and compounds injected by industry (usually illegally) into the domestic solid waste stream, sewage and stormwater. These potentially hazardous materials make it more difficult to dispose of what otherwise would be a benign mix of substances. (The problem of contamination of the solid waste stream will be developed in Chapter 5. Contamination of stormwater and sewage will be discussed in Chapters 4 and 7.)

2.2.2 Implications of waste composition

Institutions like schools, hotels and offices pose special problems and opportunities for waste management. Because they are all subject to trained management (unlike households) it should be possible to incorporate waste reduction into the management system. For schools there are important educational advantages to be gained by practising waste reduction. For all institutions there is money to be saved from waste reduction and the possibility of money to be earned by selling recyclable and reusable materials. However, in every case it requires the initiative of an individual, or a mandate from above, to put a system in place.

One of the first lessons to be learned about waste is that its composition is the key to improved management. There are three desirable attributes of waste – its potential for recycling, its potential for combustion, and its suitability for compost. Glass, paper, metals, plastics and wood can be recycled, while paper and plastics are readily combustible. Organic material can be composted.

2.3 Collection and treatment options

2.3.1 The traditional focus – waste removal

Until recently most solid waste was simply collected and buried in landfills. What was judged to be 'clean fill' could be used in construction or for landscaping, for example. Toronto has greatly extended its lakeside park system in this way, while half of downtown Tokyo stands on land reclaimed with fill. The management decisions about how to collect and bury the waste focused on the cost considerations mentioned above. Collection of the waste was the main component of cost, and labour was the major part of the collection cost. Therefore waste managers examined options such as the compaction of waste and the use of transfer stations to pool waste from smaller collecting trucks to larger trucks to haul to the landfill site. Anything that reduced the time taken to transfer waste from households to the landfill would save money. Transfer stations also offered the possibility of further reducing the volume of waste by shredding and baling.

This approach has been questioned mainly due to opposition from the residents in the vicinity of proposed new landfill sites – the NIMBY, not-in-my-backyard, syndrome. The time taken to acquire new sites and the cost of meeting new environmental and aesthetic standards pushed the cost of waste removal so high that finally planners looked for alternatives to 'collect and bury'. Previously, solid waste management responded to local taxpayers' expectations for waste removal, at the lowest financial cost. The key issue was to remove the waste from the places and activities that generated it. With NIMBY the managers had to re-examine the broader issues posed by the urban solid waste stream. Meanwhile existing landfills approached the end of their operational lives.

2.3.2 Options for reduction, reuse and recycling

The alternatives, and the responsibility for exploring those options, vary from case to case. The general goal is to reduce, or even eliminate, the solid waste stream. The means to achieve this lie in pursuing the virtuous hierarchy of reduce, reuse and recycle – a process that requires the co-operation of the waste producers (such as households and businesses) and the waste management authority, whether public or private.

For households the alternatives are shaped by their own housing situation and the service alternatives and charges offered by the authority. Households with gardens have a convenient opportunity to compost most of their kitchen and garden waste; for apartment dwellers this requires some neighbourhood co-operation, or a special service provided by the authority. Space is also a requirement for households to store bulky material for reuse (such as wood and metal) or even to store recyclable material for kerbside collection. People in apartments are usually less able and willing to turn over part of their limited space to sort and store materials.

There is no question that behaviour will change quite quickly if the right incentives are put in place, employing a combination of sticks and carrots. For example, the recent British Landfill Levy (£10 per tonne) has encouraged industries and waste management companies to reduce their flows. Builders now have an additional incentive to sort and reuse the rubble from demolition to provide the foundation for the replacement building or for landscaping. In the United States, and some European countries, households are now charged per bag of rubbish collected. Some authorities make regular kerbside collections of recyclable materials from households, especially paper, plastic, metals and glass.

2.3.3 Incineration

As the modern solid waste crisis became evident in Western cities in the 1970s municipal managers looked hopefully at incineration as a cheap alternative to 'collect and bury'. Even without any energy benefit, incineration had many attractions. It would cut costs dramatically as the destination for waste could be centralised and it would avoid the lengthy and increasingly insoluble search for new landfills in surrounding administrative areas. Truly it offered a technological escape from our predicament. The city could internalise its waste disposal problem.

The problem was that, while incineration reduced the volume of wastes, simply burning rubbish did not resolve the problem. Burning waste produced its own residuals in the form of gases and solids. The gaseous residuals were a function of the waste composition which, in modern industrial cities, included a cocktail of unknown impact. Chlorine-based plastics could release dioxins to the atmosphere, for example. Similarly, the fly ash and the bottom ash could contain a variety of potentially dangerous chemicals. These had to be landfilled somewhere.

Flue gases can be trapped, and ash can be treated to render potentially dangerous chemicals inert – at a price. The economics of incineration improves if the energy potential of burning rubbish can be harnessed.

2.3.4 Energy from waste combustion

It is estimated that a tonne of municipal solid waste (MSW) could produce 600 kWh of energy, which is useful but not overly impressive, compared with other energy sources. It puts one tonne of municipal solid waste as the equivalent of about 1.3 barrels of petroleum. Table 2.5 compares the heating value of MSW with other fuel sources. Davis and Cornwell (1998: 691) suggest that energy from municipal waste becomes viable only where both of the following conditions apply:

- Landfill is expensive (more than US$25/tonne) or unavailable if, for example, the water table is too high, as in New Orleans, and ocean dumping has been banned.
- There is a reliable, local customer and a reliable, local supply, usually requiring a population of more than 250 000 people, generating waste at the American level and composition.

Table 2.5 Net heating values of various materials (MJ/kg)

Methane*	55.5
Natural gas	53.0
Fuel oil (home heating)	45.5
Gasoline (84 octane)	48.1
Coal, anthracite	25.8
Sewage gas	21.3 to 26.6
Newsprint	18.6
Wood, pine	14.9 to 22.3
Municipal solid waste	10.5

Source: Davis and Cornwell 1998: 690.

* Densities taken as follows, in kg/m^3: methane = 0.680, natural gas = 0.756, sewage gas = 1.05.

Environmental legislation is the other key driver. If the risks from uncontrolled waste production – including greenhouse gases – become urgent then legislation to reduce them can provide opportunities for more coherent management of solid wastes. For example, in Britain the new obligation for each power company to produce 10% of its output by 2010 from non-fossil fuels has led power companies to acquire landfill sites in order to burn the waste methane to generate electricity. The fifth round of the Non-Fossil Fuel Obligation (1998) supported 141 landfill gas projects for a combined capacity of 314 MW, being 30% of the total energy produced in that round. Thus a solid waste problem has been turned into an energy opportunity.

2.4 Improving our management of the solid waste stream

2.4.1 Integrated solid waste management (ISWM)

Over the last thirty years there has been a growing appreciation of the idea that solid waste management is far more complicated than 'collect and bury'. Improvements to the system must respond to local concerns about health and amenity; they must contribute to national and global environmental objectives; they must also be cost-effective in a tightened municipal fiscal regime. The guiding premise of ISWM is a holistic linkage of the options for:

* source reduction
* recycling
* incineration
* landfilling.

2.4.2 Waste reduction

The most effective way to reduce the quantity of waste generated is to reduce the quantity of materials used, whether in the product or in its packaging. To some extent this is driven by consumer pressure to acquire the goods they want while attracting the minimum quantity of waste materials such as excess packaging. Reuse can be encouraged by legislation that requires the producers to offer only goods that are easy to reuse. Countries such as Germany have introduced laws that place responsibility on the producer for the entire cradle-to-grave life cycle of the products they make.

What is interesting is that it is rarely the *physical* problem of waste management that is difficult to resolve. For example, for households with access to gardens, household solid waste can easily be reduced by 80% once a kerbside collection system for recyclables is put in place. Participation rates are usually very high. The problem is *economic*, in that markets for recyclable materials may be difficult to establish, and the labour to handle the materials may be a significant cost, as solid waste management is a labour-intensive activity. The cost of labour also discourages consumers from repairing damaged goods; in a modern society it is often cheaper to discard and replace than to repair. The other economic problem is that as incomes rise the amount of waste generated per head rises inexorably. This is another environmental problem that awaits poor countries, as they become richer.

2.4.3 Contaminant control in the waste stream

The basic problem with the modern urban waste stream, other than its volume, is its quality. The multiplicity of industrial processes that support modern society ensures that waste streams will be contaminated. Many practices make this contamination almost inevitable. Until very recently managers of industrial processes faced only token fines if they disposed of toxics into the urban drainage system. At the same time, we commonly combine stormwater drainage with sewerage. Thus the solid residue from human waste, which is potentially reusable as fertiliser, is mixed with metals and chemicals that threaten the food supply. Furthermore, municipalities themselves used to collect commercial and industrial wastes and landfill them along with household solid waste, knowing that the industrial portion contained materials such as PVC and other chlorine-based plastics.

As long as waste management was seen simply as a process of waste removal this mixing of materials was not a problem, if landfills were physically secure. Only when you wish to reuse materials from the waste stream does the cost of this careless behaviour become apparent. Many countries are now moving towards a more selective approach to waste disposal, providing households with the opportunity to dispose of batteries, paints and solvents safely, for example. Phosphates in detergents are now being reduced. Criminal behaviour by plant managers can bring a gaol term and a significant fine. However, the problem is deeply ingrained in human activities and expectations. It is not simply a physical problem of separating the materials but of changing the entire culture of the modern 'throwaway society'.

2.4.4 Modern landfill design

In the meantime considerable progress has been made in providing better end-of-pipe treatment of solid waste by legislating much higher standards for sanitary landfills. Until legislation was introduced in Europe and North America, for example, in the 1960s, a landfill was simply a hole in the ground that was eventually covered over when it was full. This approach posed immediate costs to households in the vicinity from noise, dust, odours and a variety of chemicals that could have adverse health effects simply by inhalation or exposure to the skin. Fires in the dump intensified the dangers.

Longer term problems arose from the landfill gas which evolved as the material rotted, and from the liquid, or leachate, that began to flow from the base (Davis and Cornwell 1998: 665–9). The leachate became a problem as soon as it reached sources of the local drinking water supply. Surface water could also transport contaminants from the dump to the water supply.

Current requirements should prevent all of these dangers. First, a site must be found that cannot connect with the water supply. An impervious liner and a system of drains must be put in place to collect the leachate that is then treated. Likewise the landfill gas is collected in pipes and sometimes burned to provide energy, as noted above. The waste must be covered with clean fill daily, and finally, when full, it must be sealed to prevent subsequent transportation of contaminants.

None of these precautions makes a new landfill a welcome neighbour, nor is this approach an entirely satisfactory solution to the management of the solid waste stream.

2.4.5 Resource recovery

A goal of total resource recovery – emulating nature – is very difficult to achieve under current conditions because our goods, services and entire lifestyle in Western society are not designed for resource recovery. Goods are made, packaged and delivered without considering the ultimate fate of the materials. Only recently have some companies begun to adopt a life cycle approach to product design in which resource recovery is a guiding principle. Table 2.6 sets out various resource recovery alternatives from low technology to high.

The first steps along this road have been made more difficult by the political context in which most Western countries are currently operating,

Table 2.6 Resource conservation and recovery options

Low technology
Returnable beverage containers.
Recycling household waste based on separation by households, and collection from waste depots or from the kerbside.

Medium technology
Improved design of products and packaging.
Shredding and separation at a processing plant.
Municipal composting.
Methane recovery from landfill.

High technology
Energy from waste providing fuel to a special purpose power plant, combined with the incineration of sewage sludge, and separation of recoverable materials.

Source: Adapted from Davis and Cornwell 1998: 684–91.

namely a shift to privatisation of what used to be public services. This trend has affected water supply and solid waste management among other activities. As public budgets have been shrunk or services contracted out to private firms it has become more difficult to experiment with different approaches to waste management. Any innovation – such as recovery and recycling of materials – is now expected to pay for itself. Unfortunately, what markets do exist for recycled materials are patchy, and prices remain volatile. These economic conditions reflect the hesitancy of the legislative and political climate. Even the most environmentally progressive governments seem to take only one tentative step forward after making public declarations of commitment to a more coherent future.

2.5 Conclusion

2.5.1 Land use and waste management

Patterns of land use in the modern city encourage the wasteful use of all our resources, beginning with the land itself. We waste land by taking up peripheral farmland for construction when blighted land lies abandoned in the centre. We waste materials in the production of goods and services because our legislative system condones wasteful processes of manufacture and distribution. We then spend a lot of money collecting the solid waste and bury it at increasingly distant locations. Our land use patterns ensure that we waste energy on all the trips we make. We have developed an urban culture in which landfill dependency and automobile dependency are built-in assumptions.

2.5.2 The next steps

What can be noted here is that the first step towards improved management of the solid waste stream is to abandon the assumption of landfill dependency, the major recourse for the old mentality that said that a problem relocated was a problem solved. In extreme cases this has gone beyond simply sending wastes to a nearby landfill. Currently (February 2001), Toronto is sending 90 to 100 trucks daily to Michigan to dispose of its residential solid waste. This number is expected to double when the local landfill is closed in a few months' time. Elsewhere, toxic wastes have been shipped all over the world to find the cheapest dumping ground. It has even been proposed that the ultimate solution to the solid waste problem is to blast it into space! Fortunately this type of thinking is becoming less pervasive and is even being replaced by a realisation that used materials have an intrinsic value if handled properly.

The title of this chapter sets the goal for an ecological city. Only humans create wastes; the rest of Nature recycles everything. We began to create problems for ourselves when we congregated in cities, thereby removing ourselves from proximity to fields, rainfall and the microbial organisms that had taken care of our ancestors' bodily wastes. The nutrient resource became a health hazard, so we flushed it out through pipes to rivers and the sea. We made up the consequent nutrient deficit on the land by spreading chemical fertilisers. Similarly with solid waste, we moved from organic goods and wrappings to metals, plastics and coated paper. Most Western communities are still bundling this rubbish up, every week, and burying it in the countryside. This is a less difficult problem than the management of human sewage – it is nothing but a habit we slipped into. With sensible packaging regulations, markets for recycled materials, and urban composting (individual and communal) this solid waste stream could be reduced to zero.

In order to put ecological systems into place we need a vision of a city that could, indeed, absorb 'its own' residual materials. The last such vision was produced 100 years ago by Ebenezer Howard whose plans for a garden city included a symbiotic relationship with the surrounding farmland which would absorb residuals from the city in a beneficial fashion (Howard 1986 (1898)). Howard's plans for an ecologically sound city were based firmly on a land use plan, and also on a self-sufficient financial plan. A healthy, ecologically sound city needs nearby green space, preferably consisting of working farms, to absorb its residuals. Instead, our current patterns of urban growth have created a highly destructive interaction with rural areas. Nearly every Western city consumes the surrounding green space, while leaving dead land at its heart. It is hardly an exaggeration to describe this process as cancerous.

This process can be reversed, but it will require a conscious effort to do so. No tinkering with a few market incentives will change the direction we have taken, because the momentum to continue this mindless destruction of our natural assets is very strong. Even in a small, crowded country like Britain there are continuing plans to build houses on greenfield sites and to extend the highway system. Politicians of every stripe fail to criticise the trend for fear of being labelled 'anti-car'. Were politicians in the 1840s afraid of being labelled 'anti-disease' when they developed the Public Health Idea?

The next two chapters examine urban metabolism from the perspective of air and water, respectively. We return to land use issues in Chapter 5 when the costs of restoring wasted land and other remedial measures are analysed. Further improvements on the road to developing ecological cities are the subject of Chapter 8, 'Restoring urban land to productive use'.

2.6 Websites

1. Examples of ecological communities, based on a survey in Barton (2000):
   http://www.gaia.org
   http://www.ic.org

http://www.arcosanti.org
http://www.auroville-india.org
http://www.mcn.org/findhorn
http://www.sustain.force9.co.uk

2. Exchanges and markets for recycled materials:
   http://www.wastemanagement.com/serv_recycling.html

2.7 Further reading

Barton, H, ed (2000). *Sustainable Communities: The Potential for Eco-Neighbourhoods*. London, Earthscan.

Boyden, S, Millar, S, Newcombe, K and O'Neill, B (1981). *The Ecology of a City and its People: The Case of Hong Kong*. Canberra, Australian National University Press.

Davis, M L and Cornwell, D A (1998). *Introduction to Environmental Engineering* (third edition). Boston, McGraw-Hill.

Girardet, H (1999). *Creating Sustainable Cities*. Totnes, Green Books for the Schumacher Society. Schumacher Briefing No 2.

Howard, E (1986 (1898)). *Garden Cities of Tomorrow*. Builth Wells, Attic Books.

Jenks, M, Burton, E and Williams, K, eds (1996). *The Compact City: A Sustainable Urban Form*. London, E. and F. N. Spon.

McHarg, I L (1969). *Design with Nature*. Garden City, NY, Doubleday and Co.

Ravetz, J (2000). *City – Region 2020: Integrated Planning for a Sustainable Environment*. London, Earthscan.

Roseland, M (1998). *Toward Sustainable Communities: Resources for Citizens and Their Governments*. Gabriola Island, BC, New Society Publishers.

Wackernagel, M and Rees, W E (1996). *Our Ecological Footprint: Reducing Human Impact on the Earth*. Gabriola Island, BC, New Society Publishers.

3

Energy and emissions to the air

The changes in the atmosphere's composition provoked by human activities are unprecedented in recent geological history, extremely rapid, and – in the case of the carbon dioxide buildup – essentially irreversible.
(Harvey 2000: 8)

Contemporary London, with 7 million people, uses around 20 million tonnes of oil equivalent per year, or two super-tankers per week, discharging some 60 million tonnes of CO_2 into the atmosphere.
(Girardet 1999: 19)

3.1 Emissions to the air

The atmosphere is the most accessible sink of all, because it surrounds us and is so huge that it is very resistant to abuse. Throughout history we have discharged noxious gases which impair human health especially through the lungs, eyes and skin, but only in extreme cases have we emptied gases into it of sufficient magnitude and toxicity to make it fatally poisonous to humans. Even then the loss of life has been small compared with the deaths attributable to water-borne disease. Human discharges to the air can also create odours, reduce visibility and erode buildings and infrastructure.

In the final third of the twentieth century we have witnessed a growing awareness of the much wider impact of human, or anthropogenic, emissions through the air, such as the 'big three' issues of acid deposition, stratospheric ozone loss, and global warming (Firor 1990). Initially there was a tendency to treat each aspect of the problem separately, legislating to curb a particular gas to reduce a particular problem. Underneath these separate initiatives, however, there is an understanding that the problems have a common cause in human use of fossil fuels to produce energy, and thus could be tackled more holistically (Munn 1997; Munn *et al.* 1997). That is not to suggest that this problem area is under control. On the contrary, while in the last fifty years there has been a significant reduction, in Western cities, of particulates and emissions of sulphur dioxide to the air, this improvement has been accompanied by a continuing build-up of ozone precursors near the ground and carbon dioxide in the atmosphere. Even a windswept region like the British Isles suffers from excessive ground-level ozone, which is a contributor to respiratory problems such as asthma.

Although our use of energy from fossil fuels is the main source of these problems there are other discharges to be considered also. Rural areas contribute nitrous oxide from fertilisers, methane from paddy rice and ruminants, and carbon dioxide from forest fires and biomass decomposition. Cement factories release significant quantities of carbon dioxide. Hydrofluorocarbons (HFCs), perfluorocarbons (PFCs) and sulphur hexafluoride (SF_6) are significant greenhouse gases too, and are targeted for reduction under the Kyoto Protocol (Grubb *et al.* 1999: 300). Incinerators, mines, refineries and other factories also release a variety of gases and metals to the air. However, it is the combustion of fossil fuels to produce energy that is the major source of anthropogenic emissions to the air, and it is the management of energy use that lies most clearly within the influence of urban governments.

3.1.2 Emissions to the air from human activities

Some of the principal emissions from human activities are shown in Fig. 3.1. Their various impacts will be considered in detail in Chapter 6. What is important to note is that so many of the impacts can be traced back to the same sources. By switching from fossil fuels to a non-polluting source of energy we can win simultaneously on many fronts, from current impacts on human health to the longer range impacts of climate change, including further implications for human health. Furthermore, many of the activities that produce these gaseous emissions also produce unwanted noise and waste heat.

3.2 Cities and energy

As we saw in Chapter 2, cities are very inefficient metabolically. This is because they have evolved incrementally without any integrated planning. Now we must re-evaluate cities from the bottom up so that they can play their part in steering us away from ecological disaster. Although failure to achieve this end could be quite dramatic, the means to succeed are not dramatic at all. They simply require the accumulation of multiple small scale

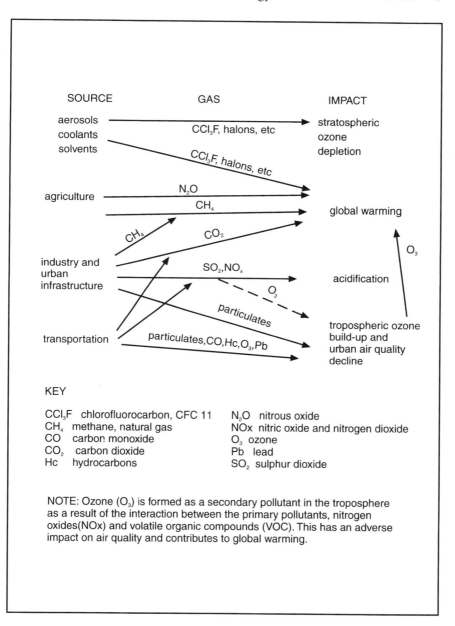

SOURCE   GAS   IMPACT

aerosols   stratospheric
coolants   CCl₃F, halons, etc   ozone
solvents   depletion
                  CCl₃F, halons, etc

agriculture   N₂O
                      CH₄   global warming
            CH₄   CO₂   O₃

industry and
urban   SO₂,NOₓ   acidification
infrastructure   O₃
                      particulates

                                                 tropospheric ozone
transportation   particulates,CO,Hc,O₃,Pb   build-up and
                                                 urban air quality
                                                 decline

KEY

CCl₃F chlorofluorocarbon, CFC 11   N₂O nitrous oxide
CH₄   methane, natural gas   NOx nitric oxide and nitrogen dioxide
CO   carbon monoxide   O₃   ozone
CO₂   carbon dioxide   Pb   lead
Hc   hydrocarbons   SO₂ sulphur dioxide

NOTE: Ozone (O₃) is formed as a secondary pollutant in the troposphere
as a result of the interaction between the primary pollutants, nitrogen
oxides(NOx) and volatile organic compounds (VOC). This has an adverse
impact on air quality and contributes to global warming.

3.1 Emissions to the air from human activities

efforts. Energy is a key to understanding the interaction between cities and
the environment on which they depend. Cities are intense concentrations
of energy use, and hence sources of residuals.

An analysis of energy use in cities tends to focus on the visible uses such
as transportation, power plants, heating, ventilation and air conditioning
(HVAC) and appliances. Yet this is only part of the energy profile that we

need to understand and modify. Before fossil fuels we relied upon the use of somatic, or bodily energy – our own, or that of domesticated animals. We also burned wood and peat, and, outside cities, we used the power of the wind for navigation, drainage and grinding cereals. Our biggest input of energy was solar energy, transformed by photosynthesis into food, both vegetable and animal. The study of this comprehensive flow of energy is known as energetics, a subject that emerged in the nineteenth century to determine whether our use of extra-somatic energy, then greatly amplified by our use of fossil fuels, was an efficient transformation of resources into food energy or not (Martinez-Alier 1990). Unfortunately, the ready availability and thermal efficiency (see Table 2.5) of fossil fuel energy discouraged a serious pursuit of this important line of enquiry.

As we look for ways to reduce energy use the question of conversion efficiency now returns. The fact is that we use a great deal of extra-somatic energy to provide the food for our increasingly urbanised society, as well as all the other products that we use. Every product that requires artificial fertilisers, industrial preparation, packaging, refrigeration and transportation can be said to *embody* the energy used in its production. Thus we should be able to assess every product for its embodied energy, be it a building, an automobile, a food product, a piece of furniture or a book. Entire industries – from tourism to television – can be assessed for the total amount of energy they require for their materials and activities, not simply the fuel they use on a day-to-day basis.

In the meantime the most profligate use of energy in the Western city is the daily use of fossil fuels that we use to keep ourselves at an equable temperature, to cook, communicate, manufacture goods and move ourselves around.

Energy use can be simply described using the R–A–R approach with an example showing energy from Metro Toronto (see Fig. 3.2). More generally the impacts of energy can be sketched to show their growing impact over time. Particulates and gases with a short residence time (generally measured in days) make their impact in and around the city. Others, like greenhouse gases, have residence times measured in decades, which is why they become thoroughly mixed in the atmosphere and hence have a global impact.

The problem with our use of extra-somatic energy is twofold – local and global. The local effects include deficient air quality that can be harmful for human health. However, these local effects are problems like the landfills and contaminated land introduced in the previous chapter, or like the water quality issues introduced in the next chapter. They have serious impacts but they can be reversed quite quickly if the political will exists to deal with them. Unfortunately, the global impacts of our energy use will require a fundamental change in our lifestyle, and even when these changes are made it will be decades before we might bring the problem of climate change under control. That is why the 'air issues' are the most critical environmental issues that we face.

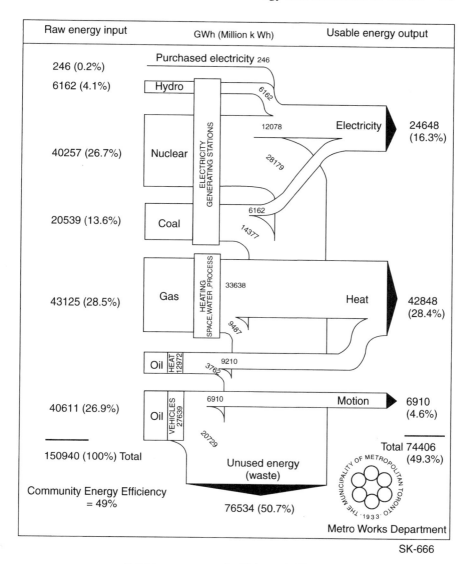

3.2 Energy use and efficiency in Toronto

3.3 Air masses and air movements

3.3.1 Altitudinal issues

Air issues oblige us to consider air masses and air movements at many spatial scales, temporal scales and altitudes. Local air quality is affected by the urban heat island that forms over every large city when air movement is low. The heat that is absorbed by buildings and streets radiates back into the air, and can raise temperatures up to 5 °C higher than the surrounding countryside. This can be an advantage in winter as it reduces demand for space heating, but in the summer it compounds air quality problems because, in direct sunlight, the higher temperature tends to favour the formation of ground-level ozone from hydrocarbons. In a world that is warming due to the enhanced greenhouse effect, air quality problems will intensify.

Altitude is important for other aspects of air quality management. Traditionally we have attempted, especially in the urban realm, to manage only the lower layers of the atmosphere, known as the troposphere, because that is the part we breathe and in which we live when we are not in an aeroplane. Ozone in the troposphere is a respiratory irritant and damages crops. In the stratosphere, the upper level of the atmosphere, is a natural layer of ozone that screens out most of the ultraviolet waveband (UV-B, 280–320 nanometres range) of incoming radiation from the sun. This protection is important for life on earth because high doses of UV-B are inimical for humans (especially for melanoma and cataracts), and for phytoplankton, the very basis of the marine food chain. Unfortunately, the CFCs we use for refrigeration, when released, ascend to the stratosphere where they break down ozone and thereby increase the amount of UV-B reaching the earth's surface. CFCs are being phased out in the Western world, but use is increasing in the poorer countries and smuggling has become a problem. Furthermore the substitutes that have been brought in – HFCs – are important greenhouse gases.

3.3.2 Regional issues

We are becoming increasingly aware of the need to manage air quality on a regional basis, identifying a local 'air-shed', which is the atmospheric equivalent of a watershed, although more difficult to bound. Ultimately this level of management will require co-operation between many authorities because problems at this scale are trans-boundary. Pollutants are now tracked from Mexico to the Arctic and from China to western Canada and the United States. Sulphates from the UK have long been recognised as the main source for acidification in Scandinavia. For Toronto, air quality is highly influenced by air moving up from the Ohio Valley, the location of major coal-burning power stations. It is estimated that more than half the quantity of pollutants found in the Great Lakes is deposited from the air.

3.3.3 Global issues

Most intractable of all is the global mixing of greenhouse gases, which results from their long residence time in the atmosphere. The enhanced greenhouse effect is the first globally distributed environmental impact attributable to human beings. Everyone contributes and everyone is affected, although some people contribute more than others, and the impacts will vary from region to region.

What is certain is that average temperatures will rise, and thus increase the probability of heat waves and droughts. The increase in heat waves is particularly ominous for cities because temperatures are already higher there than in the surrounding region because of the urban heat island effect, mentioned above. Warmer temperatures are conducive to convective storms, including thunderstorms, tornadoes and hailstorms (White and Etkin 1997). Warmer temperatures will also lead to sea level rise due to the thermal expansion of the oceans, plus whatever additional impacts result from melting glaciers and ice-caps.

Another possible, but unconfirmed, outcome from global warming is an increase in the frequency of tropical cyclones, known as typhoons and hurricanes. It is also possible that the area affected by tropical cyclones may become enlarged as the oceans warm up; this could bring major cities such as New York and Tokyo within their reach.

One of the more puzzling impacts of global warming is the effect on the movement of air masses, as it may be intensifying the El Niño Oscillation and may also be responsible for slowing down the movement of other air masses. The latter point may be more significant than it might sound. For example, both of Canada's most damaging extreme weather events owed their impacts not to the intensity of the precipitation but the duration. Both the Saguenay flood (July 1996) and the Montreal ice storm (January 1998) were the result of days of prolonged precipitation due to the fact that the responsible air masses remained stationary over the affected area. These events may not turn out to be the result of global warming but they act as a reminder that we should be prepared for 'surprises in the greenhouse'. There is no doubt that high impact weather events are becoming much more frequent, although the greater impact is partly due to the greater concentrations of population, particularly in vulnerable coastal areas.

The implications of these atmospheric changes for cities and the hydrological cycle will be explored in Chapter 6.

3.4 Energy sources

3.4.1 Conservation

As long as energy is perceived as being cheap the design of urban energy systems, such as transportation, heating, ventilation and air conditioning, will focus on other system features such as capacity or durability, with energy efficiency being a subsidiary issue. It is only recently that we have become aware of the serious side-effects of fossil fuel use (such as climate change) and of the bottom-line improvements to be gained from energy conservation (Romm 1999). It will take some time for the world to move from demand forecasting, based on extrapolation of current trends, to demand reduction through conservation. Opportunities for conservation are now being explored for every aspect of energy use in Western cities from more fuel-efficient automobiles, to domestic appliances and building design. This shift will pay dividends whatever fuel mix we rely upon in the future.

3.4.2 Fossil fuels

The combustion of fossil fuels is the major contributor towards enhanced greenhouse gas emissions (Table 3.1). This is the central technical challenge associated with building the ecological city. Emissions will be significantly reduced by greater fuel efficiency and by switching to the lower carbon fuels (from coal to oil to gas). However, to reduce the average Western citizen's annual carbon load from 3.5 tonnes to half a tonne we will have to almost

Table 3.1 Sources of greenhouse gases targeted by the Kyoto Protocol

Gas	Sources	GWP–100	% GHG
Carbon dioxide (CO_2)	fossil fuels, cement	1	81.2
Methane (CH_4)	rice, cattle, biomass burning and decay, fossil fuels	21	13.7
Nitrous oxide (N_2O)	fertilisers, fossil fuels, land conversion to agriculture	310	4.0
Hydrofluorocarbons (HFCs)	industry, refrigerants	140 to 11 700 most common is 1300	0.56
Perfluorocarbons (PFCs)	industry, aluminium, electronic and electrical industries, fire fighting, solvents	average is 6770	0.29
Sulphurhexafluoride (SF_6)	electronic and electrical industries, insulation	23 900	0.30

Source: Based on Grubb *et al.* 1999: 73.

Key:
GWP is the 'global warming potential' of each gas compared with the equivalent quantity of carbon dioxide. '100' refers to a 100-year horizon. Because each gas has a different residence time in the atmosphere the GWP varies over different time horizons. Although HFCs, PFCs and SF_6 are small in quantity they are powerful greenhouse gases; furthermore PFCs and SF_6 have atmospheric residence times of several thousand years. HFCs are growing rapidly in use as they are being substituted for ozone-depleting CFCs.

% GHG is the percentage by volume of each gas of the total greenhouse gases emitted in 1990 by Annex 1 countries, the signatories to the Kyoto Protocol.

abandon our use of fossil fuels. Given that it will be easier to make the changes in the cities rather than in smaller, scattered rural settlements, we may as well plan now for cities that can function without using fossil fuels. Contrasts in the global dependence on fossil fuels are quite striking (see Table 3.2).

3.4.3 Nuclear power

Nuclear power should be phased out too. Governments have lost the battle for public confidence both on the operating safety issue and on the safe waste disposal issue. Major users such as Germany and Sweden have signalled their abandonment of this as a power source; other countries will follow suit. Only massive government subsidies brought the nuclear power industry into being in the first place, as post-war governments tried to convince their citizens that nuclear weapons could be turned into the electrical equivalent of ploughshares. With privatisation and deregulation becoming the norm it is only a matter of time before subsidies are removed and the remaining nuclear power stations decommissioned.

3.4.4 Hydroelectric power

The potential for more electricity from hydroelectric power is fairly small in relation to the amount of energy we derive from fossil fuels. There

Table 3.2 Carbon dioxide emissions by countries grouped by income, and the six largest emitters in 1996

Countries	Total in billion tonnes		Tonnes per capita	
	1980	1996	1980	1996
Grouped				
Low income	2.1	5.1	0.9	1.5
Middle income	2.8	6.9	3.3	4.8
High income	8.7	10.7	12.3	12.3
World total	13.6	22.7	3.4*	4.0*
Six largest				
United States	4.6	5.3	20.1	20.0
China	1.5	3.4	1.5	2.8
Russian Fed	—	1.6	—	10.7
Japan	0.9	1.2	7.9	9.3
India	0.3	1.0	0.5	1.1
Germany	—	0.9	—	10.5

* World average.

Source: World Bank 2000: Table 10. Energy use and emissions, pp. 248–9.

is deepening opposition to flooding agricultural land and displacing thousands of people in order to develop what large scale hydroelectric power potential remains untapped. As glaciers shrink, glacier-fed rivers will lose some of the power they currently deliver. There is ample untapped *small* hydro potential, but most of that is located too far from cities to meet urban needs.

3.4.5 Alternative energy and renewable energy

With fossil fuels and nuclear power abandoned, and large scale hydroelectric potential quite limited, there remain only two major types of alternative. One is a breakthrough into a potentially emission-free technology such as hydrogen, or fuel cells, or a battery system for which the carbon dioxide will all be sequestered underground. These and other innovations are possibilities, although none is commercially available yet. The other major type of alternative is renewable energy. The major contenders at this time are passive solar heating, photovoltaic solar, wind power, and biomass, including methane recovery from landfills. In some regions there are opportunities to harness tidal power and wave power to produce electricity, while geothermal sources are used for heating, and groundwater can be used for cooling. The opportunities presented by renewables are examined in Chapter 9.

3.5 Energy uses

3.5.1 Transportation

The typical breakdown of urban energy by use is given in Table 3.3. Transportation commonly accounts for 20 to 30% of energy use in OECD countries, and this is the sector that continues to rise despite steady improvements in fuel efficiency and rising prices for fuel (largely due to tax increases) in some countries. For all governments – even those with better than average environmental records – this seems to be the most sensitive issue on the environmental agenda. Despite the only recent widespread availability of the private automobile and the high cost to purchase and to use, many people appear to be vehemently attached to their freedom to drive, even though this is becoming an increasingly stationary activity, especially in cities, as 'urban journeys contribute 97 per cent of the congestion costs but account for only 41 per cent of traffic' (Newbery 1995).

Only recently have there been attempts to measure the true cost of road transport (Maddison *et al.* 1996). Motorists and road haulage companies generally perceive themselves as paying far more for taxes associated with driving (fuel taxes, vehicle permits, parking fees) than they receive in benefits. Yet this is demonstrably untrue (see Table 3.4). The assumption that

3.3 Routine event. Even a moderate snowfall slows down traffic, exposing the
vulnerability of automobile dependency to the weather.
Source: Swiss Re 1994.

Table 3.3 Energy-related carbon dioxide emissions, UK, 1990–2010

End-use Sector	Carbon emissions in Millions of tonnes			
	1990	2000	2010	trend 1990–2010
Domestic	41.7	36.6	36.5	decline
Public and commercial	23.0	20.2	22.5	slight decline
Industry	48.0	44.2	47.6	slight decline
Road transport	**33.2**	**37.8**	**43.5**	**significant rise**
Other	12.4	10.9	11.7	slight decline
Total	158.3	149.7	161.8	rise

Source: Boardman *et al.* 1997: 2.

road transport is an undoubted benefit to society – despite traffic accidents,
congestion, air pollution and now climate change – is extremely strong.
For example, a detached observer would have difficulty reconciling public
outrage over a recent railway accident in Britain (in which about 30 people
died) to the widespread indifference to the annual carnage on the roads,
which averages 3000 deaths in the UK, 5000 in Canada, and 50000 in the
United States. The observer may be moved to compare this indifference to
the Mayans' reputed appetite for human sacrifice.

As outlined in Chapter 2, the key goal in this sector, for an ecological
city, is to change the modal split, shifting from dependence on automobiles
to less polluting forms of transport. If this objective cannot be achieved
then the battle to slow down global warming will be lost.

Table 3.4 UK road transport marginal external cost and road taxes, 1993 (all figures are in 1993 prices)

Externality	Cost (£ billions)
Global warming	0.1
Air pollution	19.7
Noise	2.6–3.1
Congestion costs	19.1
Road damage	1.5
Accidents	2.9–9.4
Total cost	45.9–52.9
Road taxes	16.4
Taxes as a % of cost	31%–36%

Source: Adapted from Maddison *et al.* 1996: 141, based on Department of Transport data and the authors' calculations.

3.5.2 Power generation, manufacturing and services

In stark contrast to the gloomy picture from the transportation sector there is a strengthening trend towards energy conservation in commercial buildings and factories. Even after twenty years of declining energy prices (1980–2000) many companies have achieved dramatic savings by reducing their energy demands and reducing their use of materials – a process sometimes known as 'dematerialisation'. Joseph Romm advises:

> If you are a service sector company, you should be able to cut your total carbon dioxide emissions *in half* over a five year period, with energy efficiency playing the biggest role. If you are a manufacturer, you could do the same, with at least half of the savings coming from energy efficiency and the rest from cogeneration and fuel switching. If yours is a process industry, your equipment costs the most and turns over the most slowly, but you also gain the most from the new cogeneration technology; you may require ten years to become cool. (Romm 1999: 14–15)

Some energy companies still support the lobby known as the Global Climate Coalition, which is dedicated to opposing any efforts to reduce the risks of global warming. Yet other companies in the energy sector have taken the lead in the 'decarbonization of fuel' (Chris Fay, of Shell UK Ltd, quoted in Romm 1999: 25). Shell has established subsidiary companies like Shell Hydrogen and Shell Solar (solar, biomass and forestry); it owns companies that manufacture photovoltaics and it is exploring the wind power business. Similarly, the Chief Executive of BP, John Browne, declared in 1997 that henceforth BP would reconstruct itself as an energy company, rather than a fossil fuel company (Browne 1997). More recently BP ran an advertising campaign in which 'BP' was presented as standing for 'beyond petroleum'!

3.5.3 Heating, ventilation and air conditioning

The next major users of energy are heating, ventilation and air conditioning (HVAC), lighting in households, and the commercial and industrial sector. We already have the technology to significantly reduce energy use within each building for these purposes and the payback time for the capital costs is usually quite short, two or three years being typical. (More detail on particular systems is provided in Chapter 9.) Much greater gains are possible in cities where district heating is used. For individual buildings:

> Energy-efficient systems have a higher reliability because of reduced vibration and wear and tear. They also provide greater control over air filtration, exhaust and air quality, which can improve yield as well as employee health and safety. (Romm 1999: 111)

Some building owners have found that the productivity gains from improving the work environment have surpassed the fuel savings. The energy reliability factor is likely to become even more significant if more extreme weather events make large scale power systems more vulnerable. For companies for which reliability of supply is critical, back-up generators are installed to provide uninterruptible power supply, UPS for short. There are fuel-cell systems that provide 99.9999% reliability, meaning only 32 seconds lost per year. Additional redundancy in such systems can virtually eliminate the chance of an interruption.

Given that the savings from an energy-efficient approach to business can be significant, Romm poses the obvious question: 'Why isn't everyone doing it right?' and he suggests the following reasons:

- General lack of awareness among developers, architects and engineers.
- The designers who could save clients money might find that they are paid *less* for doing so, if their fee is a percentage of the project cost.
- Developers, architects and engineers rarely perform post-project customer surveys, which in turn leads to a lack of documentation for successful case studies (Romm 1999: 76–7).

3.5.4 Household space heating, lighting, appliances and water heating

Thus it would seem that while the energy use of the transportation sector continues its unchecked rise, there is considerable room for optimism for energy conservation in the manufacturing and service sectors, as well as for the operation of buildings generally. One of the most interesting findings is that there are additional benefits, such as reliability and productivity, provided in the same package, as it were, as well as energy savings and emission reductions.

In a similar vein, research has been going on for several years to reduce energy use in households through efficiency gains (Boardman *et al.* 1997; Palmer and Boardman 1998; Fawcett *et al.* 2000). Again multiple benefits are sought, such as financial savings for the households, emission reductions to meet government commitments, and greater comfort, especially for people on marginal incomes, those who otherwise would suffer from 'fuel poverty'. This can be an acute problem for the low income elderly who may not be able to afford adequate heating in winter, or adequate cooling in summer. This is not a small matter as it is a cause of many premature deaths. In the British winter season alone this amounts to between 30000 and 60000, that is 10 to 20 times the number of deaths on the roads for the year (Boardman 1991: 149).

While space heating is the major household energy use in the UK, there is considerable scope for reducing energy use for lighting, appliances and water heating. Table 3.5 gives a detailed breakdown of a profile household energy use. It has been calculated (Boardman *et al.* 1997: 61) that behavioural change alone could reduce energy use in this sub-sector by more than 11%, by adopting such simple measures as regularly defrosting the fridge, putting lids on cooking pans and switching off the standby mechanism on the television! An important barrier to improvement is that: 'The majority of private individuals in the UK do not realise that their activities result in carbon dioxide releases, or that these are the primary cause of climate change' (Fawcett *et al.* 2000: 75). This remark complements Joseph Romm's observation concerning developers, architects and engineers, noted above. Thus, it would appear that environmental education could be a major factor in reducing atmospheric emissions and improving air quality.

Whereas certain reductions can be achieved by individual householders making informed choices, it is unlikely that significant changes will occur without coherent action by government on the regulatory front. For this Fawcett *et al.* recommend that the government should adopt a 'whole house approach' which means taking:

> the opportunity to move beyond simply the building shell and energy to specify the carbon efficiency of the dwelling. This would include consideration of primarily the water and space heating systems, but

Table 3.5 Household electricity consumption for lighting, appliances and water heating in the UK, 1998

Appliance type	GWh	% of total
Cold appliances	17492	19.9
Consumer electronics	10419	11.9
Cooking equipment	12797	14.6
Lighting	17366	19.8
Wet appliances	11845	13.5
Water heating	15406	17.6
Miscellaneous	3218	3.7
Total	88543	101

Source: Fawcett *et al.* 2000: 8.

could also include the lights and appliances installed. Whole house approaches could also contribute to encouraging fuel-switching from electricity to gas and renewable options. (Fawcett *et al.* 2000: 61)

This 'whole house approach' is consistent with the promotion of holistic approaches at the scale of the entire urban system. However, it involves a complex host of actors that differs for each sort of appliance (see Fig. 3.4).

3.6 Energy users

3.6.1 Households, businesses, governments
We can break down activities in the city either in terms of the activities themselves (as above) or by looking at the users. Each individual energy user will encompass a range of activities such as running a household, commuting to work, working in an office, going shopping and other non-work

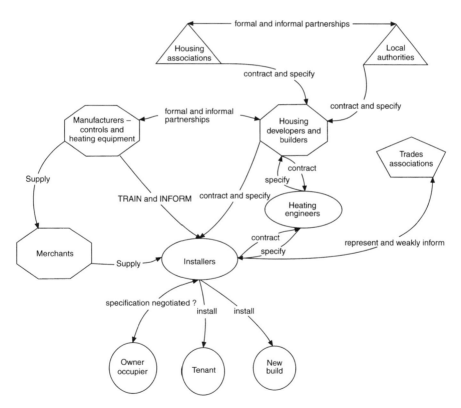

3.4 Network of actors implicated in the specification of domestic heating equipment in the UK

Source: Fawcett *et al.* 2000: 43.

activities. For each activity an energy user faces a certain set of choices. His or her choice regarding energy use may not be consistent from activity to activity. For example, an individual may strongly support energy conservation in the workplace because it obviously improves the bottom line. But the same person may not make the house more energy efficient, perhaps because the capital is not available, or it is not a priority for other household members. Likewise in transportation use, the individual is influenced by habit, convenience and sense of status as well as by monetary or environmental concerns.

In order to bring cities closer to the ecological ideal we have to understand how people behave in each of the various urban settings that they occupy. Within households, individuals may not agree on the course of action, and thus compromises are made, the most common result being inaction. Associations, such as homeowners' associations, or builders' associations, or professional associations, may take the initiative. Priorities for action will vary with each interest group.

3.6.2 Interplay between citizens, planners and urban politicians

At this time little is known about the interplay between citizens, planners, business people and urban politicians in determining an approach to something as huge as climate change. As noted earlier, it is believed that many people are still unaware of the implications of this issue, or of their central role in its evolution. However, other environmental issues have been processed through similar circumstances and lessons can be drawn from them.

The energy sector is complex and touches every facet of life in a modern Western city. Thus some variability can be expected from issue to issue. For some problems, such as fuel poverty among the elderly, it might be expected that there would be a wide measure of agreement among the various actors, even if inertia ruled the day. Transportation is probably the most difficult issue given the attachment of many individuals to their 'motorised individual transportation' by means of the automobile.

3.7 Conclusion

Strategies do exist for encouraging the 'virtuous hierarchy' of reduce, reuse and recycle. The key entry points to energy use reduction are:

* a shift in the modal split away from the automobile
* the introduction of energy efficiency as a goal in the home and the workplace
* and the switch to renewable, non-polluting forms of energy such as wind and solar power.

Energy can be reused by introducing cascading uses, such as taking waste heat from factories and power stations and pumping it into district heating and cooling systems. These strategies will be explored in detail in Chapter 9.

3.8 Websites

1. Canadian Climate Change Calculator for personal greenhouse gas emissions:
   http://www.climcalc.net

2. Cities for Climate Protection:
   http://www.iclei.org/co2/earcade.htm

3.9 Further reading

Boardman, B (1991). *Fuel Poverty: From Cold Homes to Affordable Warmth.* London, Belhaven Press.

Boardman, B, Fawcett, T, Griffin, H, Hinnells, M, Lane, K and Palmer, J (1997). *DECADE – Domestic Equipment and Carbon Dioxide Emissions. 2 MtC – Two Million Tonnes of Carbon.* Oxford, University of Oxford, Environmental Change Unit.

Boyle, G, ed (1996). *Renewable Energy: Power for a Sustainable Future.* Milton Keynes, Oxford University Press in association with the Open University.

Elliott, D (1997). *Energy, Society and Environment: Technology for a Sustainable Future.* London, Routledge.

Grubb, M (1995). *Renewable Energy Strategies for Europe. Volume 1 – Foundations and Context.* London, Royal Institute of International Affairs and Earthscan Publications.

Maddison, D, Pearce, D, Johansson, O, Calthrop, E, Litman, T and Verhoef, E (1996). *The True Cost of Road Transport.* Blueprint No 5. London, Earthscan.

Martinez-Alier, J (1990). *Ecological Economics: Energy, Environment and Society.* Oxford, Basil Blackwell.

Munn, R E, Maarouf, A and Cartmale, L (1997) 'Atmospheric change in Canada: assessing the whole as well as the parts', *Environmental Monitoring and Assessment*, 46: 1–4.

Romm, J J (1999). *Cool Companies: How the Best Companies Boost Profits and Productivity by Cutting Greenhouse Gas Emissions.* London, Earthscan.

White, R R and Etkin, D (1997). 'Climate change, extreme events and the Canadian insurance industry', *Natural Hazards*, 16: 136–63.

4

Cities and the hydrological cycle

The entire amount of water vapour in the atmosphere is completely replaced by evaporation and precipitation once every eight days. Changes in the amount of water vapour in the atmosphere will respond to changes in the rate of evaporation on a comparable time scale. Since the evaporation rate is directly related to surface temperature, it follows that the amount of water vapour in the atmosphere responds almost instantaneously to changes in the climate. (Harvey 2000: 14)

4.1 The hydrological cycle

4.1.1 Water availability

Of the major bio-geochemical cycles, water is the one we should best understand. Water is visible and we all experience it every day. Most people are well aware of the main features of the hydrological cycle, based on that everyday experience. Thus it is quite unlike other major cycles such as carbon or nitrogen, the dynamics of which remain obscure for the average citizen, whether living in a city or the countryside. People are only just beginning to understand that the carbon cycle includes the coal we burn, the trees we plant and cut down, the vehicles we drive, and the carbon dioxide we exhale. Water is quite different; we know all about water. And there is lots of it.

Unfortunately, freshwater is scarcer than most people realise. It accounts for a small portion of the water circulating through the hydrological cycle. As the human population continues to increase we are exhausting the availability of naturally occurring freshwater all over the world. In the densest industrial regions we have artificially purified and reused water since the end of the nineteenth century. Even in regions of the greatest abundance of freshwater, like the Great Lakes of North America, we have seriously polluted the water supply, such that Lake Erie was on its way to becoming biologically dead in the 1970s. Lake Baikal in Russia is poisoned by industrial pollutants. The Aral Sea has been dried out by the over-abstraction of its feeder rivers for irrigation. Lakes and wetlands are being drained around the world to provide farmland and building land.

People may think, casually, that if terrestrial sources of water become scarce then we can always tow in some icebergs (as has been done already) and invest in large scale desalination, which has already been used in water deficit locations, like Israel and Malta, for many decades. However, desalination requires energy (although some solar systems are available) and produces a lot of waste salt which is a problem to dispose of.

It is possible – as with the carbon dioxide problem – that someone will discover a technological fix and the looming problem will disappear. However, it is more prudent to assume that we have to work with what we have got, and that water is our most precious physical resource. First we need a better understanding of the local dynamics of the hydrological cycle and their relevance to the management of urban ecosystems.

4.1.2 Local dynamics of the hydrological cycle

In a world without people and their cities, rainwater soaks into the soil and pervious strata where it is stored until it seeps out to provide the base flow for streams. Storms increase the hydraulic gradient and hence the flow rate of the streams. Direct runoff occurs when soil becomes saturated or impervious surface materials are encountered. This process of precipitation, infiltration, storage, and surface flow can be viewed as an environmental service because:

> Water that infiltrates the soil supports our plants, whether crops, maintained urban landscapes, or native prairies and forests. Where we allow the system to work, water that infiltrates the soil replenishes the aquifers from which we take out well water. Its gradual discharge from the earth makes floods moderate. (Ferguson 1998: 3)

However:

> Urban development has been changing all that. Impervious pavements are collection pans that concentrate runoff and all the pollutants that accumulate on them, and propel everything immediately into streams without treatment. (Ferguson 1998: 3)

The hydrograph plots the rate of overland flow (streams, and sheet flow) in a river basin following a rainfall event. Figure 4.1 contrasts the shape of hydrographs over pervious and impervious surfaces, whether the impervious surface is natural or artificial. The faster runoff from the impervious

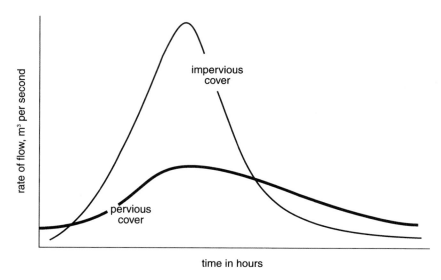

4.1 Hydrographs for pervious and impervious surfaces

surface not only increases the chances of a flood occurring, it also reduces the recharge rate of the aquifer.

Storms are characterised by intensity and duration, such as millimetres of rainfall per hour. Data are collected to record storms over time for any given location to produce intensity–duration–frequency curves. Graphs of these IDF curves are then used to design urban infrastructure to withstand a particular 'design storm' with a given 'recurrence interval' or 'return period', such as the 10-year storm (see Fig. 4.2).

The concept of a 'design storm' is a basic tool for managing runoff to reduce the danger of floods. In order to understand the hydrological cycle in a way that is useful for ensuring an adequate supply of water, a more aggregated concept is required. This is provided by the concept of the 'water balance', which like the design storm is specific to a given locality. The water balance has been described as:

> a complete inventory of the hydrology of a landscape. It is complete in the sense that it is extended over time, aggregating the effects of many small and large storms and showing the changes with the seasons. It is complete also in the sense that it characterizes the entire regime of where water is and what it is doing; it presents a unified view of the moisture environment. (Ferguson 1998: 85)

The basic concept, developed by C W Thornthwaite in the middle of the last century, is simplicity itself (Fig. 4.3) but the data requirements to make it operational are rather more formidable, while its relevance for urban design decreases with seasonal and annual variability. Generally the basic data for a water balance model are measured at a site – a particular point on

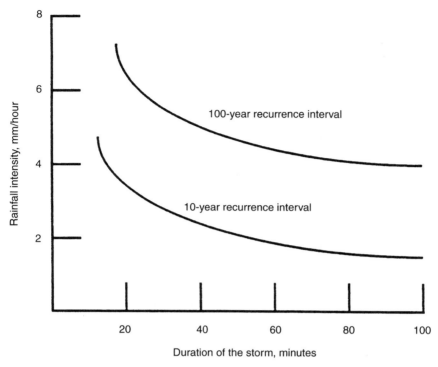

4.2 Typical intensity–duration–frequency curves

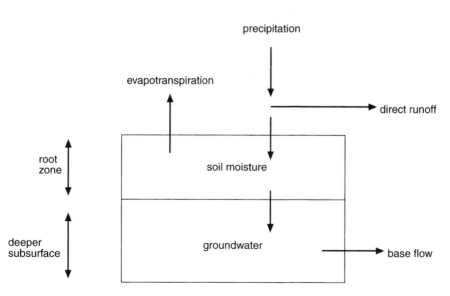

4.3 Simple water balance model
Source: after Thornthwaite 1955.

the map. The understanding of the water balance for a river basin requires a sample of sites that will adequately characterise the basin's hydrology. It becomes more complicated to apply the concept of the water balance to an urban ecosystem because an urban area rarely corresponds to a simple set of physiographical units such as a series of adjacent drainage basins.

4.2 Urbanisation and water use

In a typical modern city, water usage is partitioned between households, industry and commercial and institutions, roughly according to the breakdown illustrated by Table 4.1. Some urban jurisdictions also include significant agricultural use, as is the case for Beijing where irrigated agriculture is a major factor in water demand. The interplay between agricultural and urban uses in the same catchment area can become dangerous when agricultural fertilisers and biocides contaminate the drinking water, and, conversely, urban–industrial users pollute aquifers from which contaminants are taken up by plants.

In most Western cities water has been provided very cheaply, often at a fixed charge for households, and with discounts for bulk users in industry and institutions. Even when usage is metered the cost is very low. In Canada a typical household pays less than a dollar (approximately 40p) for the use and treatment of a cubic metre of drinking water; in Britain the equivalent charge is just less than a pound. A cubic metre will provide for the daily needs of a household of four to five people.

The reasons for providing water at bargain prices are historical and were valid for their time. In the 1840s, when the Public Health Idea resurfaced in Western Europe for the first time since the Roman Empire, clean water was needed desperately to offset the growing threat from cholera and typhoid that flourished in the cramped living conditions produced by the early phases of the industrial revolution (Goubert 1989). It was provided at low cost because it was urgently needed as a public good to ensure that industrialisation – located in dense urban settlements – could proceed. Indeed it is likely that a clean water supply is the single most important

Table 4.1 Sectoral water withdrawals by country income group

Country income group	Annual withdrawal per capita m^3	Withdrawals by sector (%)		
		Domestic	Industry	Agriculture
Low income	386	4	5	91
Middle income	453	13	18	69
High income	1167	14	47	39

Source: World Bank 1992.

factor in reducing infant mortality rates – the most sensitive indicator of environmental quality. Later, water was supplied at low cost to commercial and industrial activities in urban areas to attract jobs. Today the same arguments apply to some extent. An employer in a water-intensive industry like brewing, faced with escalating water costs, may choose to relocate.

The problem is that 150 years of low cost water have blunted popular appreciation of the vital need that water fulfils. It has become the largest component of urban metabolism by far (see Fig. 1.3 and Table 2.1). In water-scarce regions an individual might survive on as little as 30 litres of water per day, for drinking, cooking and washing. In a modern city, 200 litres per person per day is the low end of the usage scale. In urban areas, in suburbs provided with gardens and automobiles, daily personal usage of 600 litres is common. While it is true that the dirtier atmosphere in cities might increase demand for washing people and their clothing, this discrepancy is not due to need, but to habit. In rich countries people water their gardens and wash their cars without considering the cost of water because they pay so little for it, and often the rates are still fixed as a lump sum. Urban people use more water than rural people simply because water is cheaply available.

Water use rises with income, just like the use of most resources. The sectoral breakdown of use also shifts with rising income, as a smaller percentage of withdrawals goes to agriculture, in favour of a rising take from industry and from households (see Table 4.1).

As world population rises and urban systems grow larger this profligate use of water has become difficult to sustain, both for reasons of the physical availability of raw water and because the costs of provision are often met by cash-strapped local authorities. The logic for metering all users and charging enough to cover all aspects of supply – including capital costs – is becoming more widely appreciated. This change has been accompanied by the privatisation of the water industry that is supposed to provide the necessary capital to ensure a sustainable water supply. Water supply has since become an international affair with companies bidding to take over water operations around the world. Providing public bodies (such as the Drinking Water Inspectorate in the UK) can regulate quality, service and price, this development is probably a useful step towards facing the mounting water crisis.

As with energy use, we can look for additional means of supply but the key strategy, in the rich cities of the world, is to waste less. We will begin to do this only when we monitor what we use and pay the full cost of provision.

4.3 Urban impacts on the hydrological cycle

Humans living in cities use more water than their rural relatives do and they also introduce a more diverse set of pollutants into the waste stream, due to the complex range of industries located in urban areas. Cities – as a constructed environment – directly affect the flows of the hydrological cycle into which they insert themselves. As noted by Bruce Ferguson:

> site developments, of all kinds, involve impervious and compacted surfaces. The change in land cover increases runoff over the surface, dumps flood waters into streams, reduces groundwater recharge, diverts water from base flows, and turns oil from the streets into pollutants. (Ferguson 1998: ix)

The most direct impact is the effect of paving over what used to be porous surfaces (such as farms, forest, river banks and gardens), thereby preventing entry of precipitation into the ground. Smaller streams are also covered over, and some are converted into sewers. Of the rivers that do remain flowing through a city the quality is not good. The judgement of Ferguson, writing of the United States, is widely applicable:

> Contemporary urban streams are characterized by high bacterial density, high oxygen demand, high concentrations of solids and nutrients, high turbidity, and high concentrations of metals and compounds. The numbers and diversity of fish are low, and they decline as the impervious coverage in the drainage area increases. (Ferguson 1998: 9)

Reduction of the porous ground surface and surface waters, significantly increases the rate of precipitation runoff, thereby increasing the risk of flooding. This can lead to serious consequences in cities that use combined sewerage systems where the stormwater flows through the sanitary sewers. As the amount of porous urban land decreases, the likelihood of the water load exceeding the pipe capacity (or the treatment capacity of the water treatment plant) increases. When this happens untreated sewage is released to the nearest available water body. Effectively what urbanisation has done is to increase the likelihood, or 'return period', of floods in urban areas. This is an inadvertent impact on the hydrological cycle, in addition to the deliberate impacts of urban activities such as water used for cooling power stations and industry, and water used directly by households or industrial processes. Works are also carried out to contain natural stream flows by replacing them with concrete channels.

Urban systems also abstract significant quantities of water directly from surface water, by pumping groundwater, and by creating surface reservoirs in the surrounding catchment areas. If these activities are not carried out in a sustainable way they can have a number of adverse effects, the implications of which are detailed in Chapter 7. Over-pumping of the groundwater can lead to subsidence, as described in Chapter 5, section 5.6. Abstraction from surface water can reduce stream flow, which may affect downstream activities, such as other human water users and fish. Storage in surface reservoirs increases evaporation; in hot climates this can lead to salinisation of water remaining in the reservoir. Probably the most serious impact on the natural cycle is the quantity and variety of pollutants that urban activities introduce.

In summary, urbanisation, under current practice, has the following impacts on the natural hydrological cycle:

- It increases the importance of direct runoff relative to infiltration, and hence increases the risk of flood.
- More water is abstracted for human use, as incomes increase and the population becomes more urban.
- If abstraction exceeds aquifer recharge then the land may experience subsidence.
- A variety of contaminants, both pathogenic and toxic, are introduced to the water.
- These impacts combine to reduce the variety and abundance of aquatic fauna and flora.

4.4 Urbanisation and water management

The urban water manager is thus faced with three principal challenges:

- to reduce the risk of flood
- to ensure the amount of water required for human use
- to control contaminants in the water supply, most of them introduced or amplified by human activities.

4.4.1 Reducing the risk of flood

There are two routes immediately available for direct runoff – conveyance, and detention. 'Conveyance' by stream, by pipe or by swale – constructed, or vegetated – refers to the gravity-fed movement of water through the urban system within the hydrological basin. Swales are 'open channels with unobstructed flow' and 'Until about 1965 conveyance was the exclusive stormwater approach in almost all urban areas of the United States. It got rid of on-site nuisances' (Ferguson 1998: 113).

This approach is quite similar to the old approach to solid waste management – get it out of the city! As in the waste situation, we are now seeing a more fundamental appraisal of the issue. The flood risk has itself been greatly intensified by the traditional approach to stormwater management – move it through the city as quickly as possible, to minimise the risk of build-up and flood. By contrast, more attention is being paid to the potential for detention – holding the water in basins, and slowly releasing it to the stream while giving it the opportunity to infiltrate the soil. Reconsideration of the problem has led to the development of a whole range of 'green infrastructure' options, which will be considered in detail in Chapter 10 (section 10.3).

4.4.2 Ensuring the supply of water

Once again, practice is changing. In the past, demand projections were made and additional supplies were appropriated. For most cities in the last 150 years this has simply meant pulling in surface supplies from further and further afield, if additional groundwater sources were not available. As long as the population and their cities were small this approach was probably justifiable. Now it no longer works. Even in countries like the UK with regular rainfall, or the United States with massive rivers, supplies have been pushed to their limits. As with energy and solid waste, managers are now trying to 'manage' – meaning reduce – demand, rather than automatically increase capacity.

It is important to be conscious of how close to the edge advanced industrial countries had come in the immediate post-war period when economic growth became an end in itself, instead of a means to an end. When local drainage basins became 'fully allocated' the next obvious step was to move into the next basin and set up an inter-basin waterway to feed the deficit area. Among the proponents of such schemes there was little concern about the environmental implications of reversing the flow of major rivers, such as Russia's northward flowing rivers to irrigate the parched south. Various forces have checked this tendency in America and Russia, but in China the spirit of the massive inter-basin transfer solution is still alive, with plans to divert major flows from the Yangtze in the south to the Huang Ho in the north.

4.4.3 Controlling contaminants

Supplying more water, without the additional capacity to treat it, is actually a threat to human health, as the untreated water, once returned to the hydrological cycle, is a breeding ground for bacteria such as *E. coli*, for disease vectors like mosquitoes, and a medium for contaminants like phosphorus which encourage eutrophication. In addition to the threat of disease we have to deal with the other contaminants that we have introduced into the environment, such as synthetic chemicals, and higher concentrations of metals and nutrients. Some of the nutrient overload comes downstream from agricultural activities, while more is added by the use of phosphates in detergents.

As population density increases all these threats become more acute. At the same time the physical availability of water becomes scarcer. Both these factors require us to treat all used water to return it to potable condition. At a certain population density and land availability this can be achieved naturally by using mechanical filters, settlement tanks and microbial digestion, before returning water to the ground. With more intensive use the water may require tertiary treatment and may need to be sterilised by using chlorine, for example. However, at certain concentrations chlorine can be carcinogenic, so there are limits to its application.

In addition to the contamination threat from sewage there are also the chemicals introduced into the cycle from industrial activities, including metals like cadmium, arsenic, mercury, lead, chromium, zinc and aluminium, as well as compounds from solvents, PCBs and furans. In theory all these wastes should be treated by the producer; however, in practice, all kinds of pollutants are put into the drains, and hence mix with the sewage and reappear as contaminants in the sewage residue. This, in turn, makes it dangerous to incinerate sewage sludge, or to return it to the fields as fertiliser.

As long as people and their associated activities remain small in relation to the regional availability of freshwater these contaminants may not be a problem. Beyond that level constant treatment and monitoring is essential to maintain water quality. Results from an annual inspection of a water supply company in the UK are shown in Table 4.2. Routine tests cover pathogens, pesticides, minerals and by-products of industrial processes, as well as factors that influence the water's acceptability to the consumer, such as odour, taste and colour. Other substances of concern are added to the inspection list as and when directed by a higher authority (such as the European Union for the UK) or when they are detected as a problem locally. For example, Britain is bringing in new tests to detect the protozoan *Cryptosporidium*, which is discussed in more detail in Chapter 7.

The maintenance of this kind of standard becomes more difficult as the population and its activities increase in size and complexity. For example, the computer industry may not emit pollutants to the air like traditional industry but it introduces new sources of pollutants such as chromium, cadmium, silicon and mercury. Cities like Beijing industrialised under a communist regime for which environmental standards were non-existent; now it must follow a painful and expensive path to restoring water quality. Further uncertainties for both water quality and water quantity are posed by climate change. (These issues will be taken up in Chapter 7, 'The water we use and abuse'.)

4.5 Climate, climate change and water supply

The availability of urban water in acceptable quantity and quality is highly dependent on climate. The most significant variable is the total precipitation falling in the catchment areas on which a given city draws, both from surface water and from the ground. The seasonality and reliability of the rainfall is the next most important factor – the more seasonally concentrated the precipitation the greater the need for storage capacity. For example, Britain is blessed with a fairly evenly distributed rainfall throughout the year, which partly compensates for the small area of the watersheds.

Temperature is the next critical factor. In climates with significant snowfall accumulation through the winter, the spring thaw will cause extensive

Table 4.2 Summary of water quality tests made on Thames Water Utilities Ltd for the UK Drinking Water Inspectorate in 1998

Parameter	Total tests	Tests contravening the standards	
		Number	%
Coliforms	19515	325	1.7
Faecal coliforms	19515	13	0.1
Colour	2098	0	0.0
Turbidity	2124	0	0.0
Odour	1214	2	0.2
Taste	1195	1	0.1
Hydrogen ion	2099	0	0.0
Nitrate	2120	2	0.1
Nitrite	6803	193	2.8
Aluminium	2123	0	0.0
Iron	2543	16	0.6
Manganese	2095	0	0.0
Lead	1065	3	0.3
PAH[1]	1284	28	2.2
Trihalomethanes	955	1	0.1
Total pesticides	15006	0	0.0
Atrazine	14526	10	0.1
Diuron	14526	0	0.0
Isoproturon	14526	0	0.0
MCPA[2]	7708	2	<0.1
Other pesticides	201875	0	0.0
Ammonium	2120	0	0.0
Benzo-3,4-pyrene	1284	2	0.2
Copper	1065	0	0.0
Mercury	259	2	0.8
Oxidisability	236	0	0.0
Potassium	242	1	0.4
Sodium	242	1	0.4
Zinc	1065	0	0.0
All others	37727	0	0.0
Total	379155	602	0.2

Source: UK Drinking Water Inspectorate, available at *http://www.dwi.detr.gov.uk/pubs/coreport/tha98.htm*. Crown copyright.

Notes:
1. Polycyclic aromatic hydrocarbons.
2. A herbicide.

The report concluded, 'Thames Water continues to supply you with drinking water of a high quality. In 1998, 99.84% of more than 454200 tests met the standards. This is above the overall figure for England and Wales of 99.7%' (Drinking Water Inspectorate 2000: 1).

floods that may affect urban areas and their water quality. An additional source of contamination may arise from large quantities of salt used through the winter to keep highways clear of snow and ice. Conversely, warmer climates provide a more benign environment for pathogens and water-related health issues are consequently more difficult to manage.

Climate change will affect all the above factors. The hydrological implications of climate change were summarised by the Intergovernmental Panel on Climate Change as follows:

> Warmer temperatures will lead to a more vigorous hydrological cycle; this translates into prospects for more severe droughts and/or floods in some places and less severe droughts and/or floods in other places. Several models indicate an increase in precipitation intensity, suggesting a possibility for more extreme rainfall events. Knowledge is currently insufficient to say whether there will be any changes in the occurrence or geographical distribution of severe storms, e.g. tropical cyclones. (Houghton *et al.* 1996: 7)

The most critically affected cities will be those that occupy coastal locations close to sea level (Timmerman and White 1997). For example, half of

4.4 A recurrent problem. The slow subsidence of Venice, due to the over-pumping of groundwater on the nearby mainland, makes it very vulnerable to floods. Walk ways across the Piazza San Marco are a routine occurrence.
Source: Munich Re. 1997.

metropolitan Tokyo is less than one metre above sea level, and half of that area actually lies below the high water mark. Cities should expect more flooding directly from the sea, and from back-up of the water flowing down-stream. For coastal cities that are already sinking due to the over-pumping of groundwater (such as Bangkok and Venice), or due to isostatic re-adjustment (as is the case for London), this is a serious combination of problems. London – like Paris – is also suffering from a rising water table as a decline in industrial withdrawals permits a replenishment of the aquifer. There will be additional problems for coastal cities due to back-up of sewage outlets and to discharge from sewage treatment plants. Salt-water intrusion of the aquifer may also be a problem for some.

The changing distribution of rainfall will require adjustment also. In south-east England, for example, climate change is expected to bring drier summers and wetter winters, which increases the risk of shortages in the summer and floods in the winter (Wade *et al.* 1999). An increase in seasonality will usually mean that there will have to be more investment in water storage facilities. The increased intensity of rainstorms will also require increased floodwater retention.

Urban water management is already heavily dependent on climate and is under pressure from growing population and industrial demand. Response to these challenges will be made more difficult by the additional impacts of climate change.

4.6 Conclusion

Even before the threat of climate change it was becoming obvious to many urban planners, managers and environmentalists that the old ways of urban water management had to change. Stormwater could not be simply flushed through the city to the nearest watercourses because the contaminants in the stream, especially in the 'first flush' after a rainstorm, created health problems downstream, reduced amenity value, and required costly clean-up. The city has to learn to play its part in a more natural hydrological cycle, retaining some of the floodwater by detention and allowing more to infiltrate into the soil and the aquifer. Our cities have to become more porous through the reintroduction of green space and the building of porous infrastructure.

Similarly, projected supply shortages could not simply be met by drawing in more water from distant basins or by mining groundwater. The costs of these unsustainable approaches were becoming clear even in regions that were blessed with generous supplies of water. It is better to curb water use and carefully reuse what we have. An important part of reuse is to make sure that contaminants are excluded from the stream as rigorously as possible. In the past this has been a low priority because it is difficult to enforce and the public still believes that water is cheap and plentiful. Like the price of energy and solid waste management, the price of water will probably have to rise to provide the capital to meet the costs of designing to a higher standard.

None of this will be easy because these issues have been only intermittently problematic in the past. A drought could be attributed to unusually low rainfall, not poor management. A flood could be blamed on unusually heavy rainfall, not on an outdated approach to stormwater management. Even without climate change there is an urgent need to reassess – and reposition – the city in the natural hydrological cycle. The uncertainties introduced by climate change only make this need more urgent.

4.7 Websites

1. Foundation for Water Research: *www.fwr.org*

2. New York City Water Supply Report 1999: *http://www.ci.nyc.ny.us/html/dep/html/wsstate.html*

3. Thames Water: *www.thames-water.com*

4. The World's Water – Information on the world's freshwater resources: http://www.worldwater.org/links.htm

5. Toronto Water Supply Report 1999/2000: *http://www.city.toronto.on.ca/water/*

6. UK Drinking Water Inspectorate: *www.dwi.detr.gov.uk/h2oinfo.htm*

7. United States Environmental Protection Agency, Office of Ground Water and Drinking Water: *http://www.epa.gov/OGWDW/*

4.8 Further reading

Davis, M L and Cornwell, D A (1998). *Introduction to Environmental Engineering* (third edition). Boston, McGraw-Hill.

Ferguson, B K (1998). *Introduction to Stormwater: Concept, Purpose and Design.* New York, Wiley.

Goubert, J-P (1989). *The Conquest of Water: The Advent of Health in the Industrial Age.* Cambridge, Polity Press.

Hough, M (1995). *Cities and Natural Process.* London, Routledge.

Timmerman, P and White, R R (1997). 'Megahydropolis: coastal cities in the context of global environmental change', *Global Environmental Change*, 7, 3: 205–34.

Wade, S, Hossell, J, Hough, M and Fenn, C, eds (1999). *The Impacts of Climate Change in the South East: Technical Report.* Epsom, W S Atkins Ltd.

Part III

Pathology: what's gone wrong?

5

Urban land: asset or liability?

*In nature there is no waste because, after use, materials are
re-assimilated into the landscape by the prodigious work of
micro-organisms. Instead of an industrial pattern of one-way flows from
source to use to waste sink, natural systems are based on cyclical
material flows from source to use to source. . . . Thus nature's pattern is
regenerative, or inherently self-renewing, in contrast to the degenerative
industrial pattern, which is inherently self-destructive. (Lyle 1999: 155)*

*During the last 150 years the effects of industrialisation have ruptured
the equilibrium which by recycling natural wastes had remained in
balance over a long period of time. The sudden discharge of masses of
toxic residues into the natural milieu has gradually led to numerous new
risks to the equilibrium of the environment and ecosystem, as well as for
man himself, the producer of these wastes and the causative agent of this
monumental disequilibrium. (Lecomte 1999: 3)*

5.1 Paying for the past

As the richer countries of the world devote an increasing proportion of
their assets to re-establishing urban environmental quality they must con-
front a paradox. They are paying for mistakes made by previous generations
in ignorance of the long term impact of their actions. Few citizens of those
countries are aware of this historical debt on their shoulders. The paradox
is that they are equally unaware that they are leaving an even greater debt
to their descendants who will have to deal with the resource scarcities
(water, soils, forests, etc) and climatic disturbances that are the direct result

of our current actions. Perhaps we will grow to understand the role of our generation in the unfolding of environmental history; in the meantime we have to pay for the past actions of our recent ancestors.

One segment of society that is learning about this burden of the past is the property and casualty insurance industry. According to the law, insurers may be responsible to meet current standards on claims against old policies. Invariably, these standards are higher than those of a hundred years ago, the heyday of the heavy industry phase of the industrial revolution. This situation is known as retrospective liability, or paying for the past.

The major triggers for claims against old policies relate to contaminated urban land, landfill sites, fuel storage tanks (especially underground tanks), accidental spillage of contaminants, lead paint, and asbestos. In the insurance industry these are sometimes grouped together as 'asbestos and environmental risk'. This threat seemed to appear out of nowhere, as noted in an industry report:

> there is still an inadequate understanding by insurers of the industry's *potentially enormous liability* for cleaning up past environmental damage. In the United States, for example, a recent report by A. M. Best Company estimates the industry's potential exposure to environmental and asbestos claims over the next 25 years at $132 billion, or 72% of insurers' current total capital and surplus.
> (Insurance Bureau of Canada 1994: 24)

Such claims were unanticipated by the industry; thus reserves were not set aside to meet them. Some companies have been forced into bankruptcy, usually while paying the legal bills to defend themselves. Lloyds of London hovered on the brink as new claims poured in during the late 1980s following a major expansion of the group into the United States, just as claims were gathering momentum in the courts. The company estimated its losses for 1988–92 as £7.9 billion.

Apart from the legal and ethical dimensions of this process, there are serious problems in gathering the data necessary to deal with claims from old policies. Sometimes insurers cannot find a copy of the relevant policy due to takeovers, relocations and accidents such as fires and floods over several decades. Municipal records may be imprecise as to land ownership, land use and the location of underground infrastructure. Even critical definitions as to what constitutes 'a contaminant', for example, are not agreed upon, and they vary from one jurisdiction to another. This upgrading of environmental standards – applied retrospectively – has had unforeseen consequences for environmental management. For example, the risks associated with contaminated land may now mean that land that was once regarded as a valuable asset may become a major liability for owners, developers, lenders and insurers.

This problem is part of the inherent contradiction of the urban dynamic. People concentrate in urban centres to provide mutual accessibility to goods and services. If they do not manage their wastes effectively, however, they risk poisoning the most central, and hence most valuable, parts of the urban system they have created. And it is undeniable that we have *never* managed our urban wastes entirely effectively. Since the mid-nineteenth century we have dealt with the problem of accumulated wastes either by exporting them to the countryside, or by selectively abandoning industrial land that has become too contaminated to continue in beneficial use. A proliferation of landfills (rubbish dumps) and urban blight are the twin symptoms of this system failure.

Industrial technology has allowed us to continue to become materially richer while postponing payment for running down our biospheric resources. As that technology has evolved it has moved further and further from dependence on organic materials (which eventually would be assimilated by natural processes), to inorganics like brick, steel and cement. In the twentieth century we began to use synthetic materials, like nylon, PCBs, dioxins and furans; we have amplified major bio-geochemical cycles such as phosphates and nitrogen; and most recently we have begun the genetic modification of plants and animals. The impact of these substances and activities is largely unknown. The travels of materials through the environment are quite difficult to trace. Their impacts on humans and other animals are not always known either. The development of nuclear power belongs to this same process of innovation and use, without understanding the wider implications. The fact is that humankind, for all its much vaunted ability to reflect, is – from a biospheric perspective – a very unreflective, short term thinker.

Many of the problems associated with these past actions have now come home to roost.

5.2 Contaminated land and urban blight

Empty lots in city centres are a common feature of many urban landscapes. They may be empty while waiting for conversion to a new use, perhaps caught in an economic downturn and used for parking until demand for office space picks up. Like unemployment, our society needs a certain amount of unused land to provide fluidity in a dynamic economic system. However, in the last twenty years Western cities in their post-industrial phase have found that contaminated land has become a threat to urban quality of life. This problem arises from the conjunction of two quite separate processes. One process is the improvement in detection standards for contaminants left in the ground and the knowledge that contaminants above certain levels of concentrations may have serious health effects for humans. For example, mercury poisoning can have serious impacts even on adults, as demonstrated by the Minamata case. The effect of lead on children will also be examined in this chapter. Other contaminants have been observed to have serious adverse impacts on animals and thus there is concern that they may also affect humans.

The other process is the increasing pressure on land use from the continuing growth of cities. The abandonment of contaminated sites and continued economic growth mean that farmland (known, in the construction industry, as a greenfield site) is converted to urban use. This increases the size of the city and hence may lengthen the average journey to work and other journeys in the urban system. The removal of peripheral green space also reduces amenity levels on the periphery and adds stress to the quality of air and water for the city and surrounding area.

Until fairly recently few citizens were concerned about these trends. Most new homeowners who could afford the choice preferred new, low density homes in the suburbs. If new jobs also went to the suburbs then the suburbanite's journey to work might avoid the city centre altogether. Cities 'hollowed out' as automobile ownership continued to rise in the post-war West, and more people moved to the suburbs. In many parts of the United States the city centre came to be occupied only by civic buildings, prestige office blocks and parking lots. At night it was deserted. In the larger cities many people living in the suburbs ceased to visit the urban core altogether.

This process might have continued unabated, eventually affecting all growing cities that were wealthy enough to support it. However, a significant number of people realised that this trend was highly wasteful of resources, and, in the idiom of the late 1980s, it was not sustainable. The abandonment of old urban land was a straight loss of an asset, and its idle state had a blighting effect on the surrounding area. The blight further encouraged those residents who had a choice to move out to the suburbs. Blighted lands have been the target of urban renewal schemes in the West throughout the post-war period, with only limited success. Some owners of old urban sites simply abandoned them to their creditors if no beneficial use could be found. The banks began building up a portfolio of useless property, which in turn became a significant risk for them to bear. As observed by Phil Case:

Contamination can give rise to . . . indirect risks for a lender i.e.

- how the cost of cleaning up can weaken a borrower's covenant to repay a bank; and /or
- how a diminution in the value of land/property can weaken a company's balance sheet (perhaps thereby tripping a facility letter covenant linked to net tangible assets); and/or
- how security values can be adversely affected by contamination leading to reduced recovery of debt on the sale of the security – or worse still, no recovery at all.

Historically, the most visible (to a bank) of the three risks outlined above has undoubtedly been the third. Major financial institutions in the USA, Switzerland, Germany and the UK have each suffered losses counted in tens of millions of dollars as a result of contaminated security. (Case 1999: 109)

In the meantime some jurisdictions, especially in Europe, became periodically concerned about the loss of green space in the surrounding countryside. In regions like Holland the scarcity was visibly apparent to nearly everyone and further expansion was clearly coming to an end. Thus the question arose: how can idle urban lands be returned to beneficial use and the remaining countryside be preserved?

5.2.1 Standards

There is no obvious answer to this very simple question. Modern society embodies many contradictory expectations, having gone in a hundred years from total disregard for pollution to one of deep concern. Every modern jurisdiction has standards for air quality and water quality, but the approach to polluted land is less coherent. For some, all contaminants simply have to be removed, even though there is little agreement on what constitutes a 'contaminant', other than an assumption that a contaminant is a substance or organism that might pose a threat to human health. There is little consensus on the levels of contamination that might be dangerous. In general, the concentrations of concern have become lower and lower as detection levels have improved. Background, or natural, concentrations for a substance might be considered 'safe', but there are no background levels for synthetic chemicals, or for by-products of hydrocarbons. Everyone agrees that contaminated sites should be cleaned up before being put back into use; but there are no ready answers to the question: how clean is clean?

Initially government authorities wanted the removal of all contaminated land to a 'safe landfill' and the replacement of that earth with clean fill. Some authorities wanted earth replaced right down to the bedrock! A major problem compared with air and water standards is that the earth is much more difficult to sample for contaminants because foreign substances in the soil are less visible and less evenly distributed than they are in air and water. A site could be sampled, and duly remediated, only to find during clean-up or construction that the pollution was much worse than the sample had suggested. Another problem lies in the unpredictability of the underground transport of pollutants through groundwater and the risk of migration of pollutants off the site. Again, tests can be run, but they never reveal the full complexity of the flows.

To make matters worse, some jurisdictions do not require the vendor to reveal the presence of known contaminants. Despite the possible risk to public health, in some countries a sale may proceed on the basis of *caveat emptor* – buyer beware.

For both technical and political reasons, progress on the reuse of contaminated urban land has been slow. In certain jurisdictions it has been accepted that remediation standards should be site-specific, not universal, because the potential health risk is a function of the use to which the land is to be put, as well as the nature of the contaminants in the ground. A site would need to be cleaner for use as a park, school or urban garden than for use as a warehouse or underground parking lot.

It must also be recognised, however, that although our urban view of the

Table 5.1 Examples of potential dangers as a function of activities executed on or around a polluted site

Usage	Potential dangers to a population
Domestic gardens	Toxic effects due to direct contact with soil.
Groundwater for drinking	Contamination of water sources. Toxic effects due to direct contact with water. Ingestion.
Playgrounds and other public infrastructure	Direct contact. Corrosion of underground structures, such as sewers and storage tanks.
Buildings for storage, production & commerce	Release of gases. Contamination of drainage networks.
Underground structures for storage and transport	Fire, explosions.

Source: Adapted from Lecomte 1999: 50.

problem of contaminated land is focused on specific sites, the impacts may be much wider because contaminants may be widely dispersed by air and by ground- and surface water. The target population extends beyond the actual users of the site to the wider public who may breathe or drink the contaminants.

5.2.2 Techniques for remediation
The principal options for the treatment of contaminants are:

- physical removal (for treatment off site)
- physical containment *in situ*
- chemical and electro-chemical treatment
- thermal treatment
- biological treatment (Lecomte 1999, Chapter 5 *passim*).

Selection of the appropriate method is a function of several factors, both physical and socio-economic. Physical factors include the nature, concentration and abundance of the contaminant, the nature of the media (soil, water) in which it resides, the dynamics of the hydrogeology (including climatic factors), and the space available on site for treatment. The socio-economic issues are agreement on ownership and liability, whether or not the site is in use, if neighbouring lots are occupied, the cost of remediation alternatives, and the time required for treatment. Some techniques, like pumping and venting, may take two or three years to achieve the desired result.

Remediation techniques are most efficacious when there is only one contaminant, where the contaminant has been recently introduced, where the hydrogeology is well understood (i.e. quite simple), and where the medium (soils and water) is fairly homogeneous. Unfortunately these conditions are never met on an urban site. The history of the site is usually incomplete, the contaminants are mixed, and even though the natural medium may have been homogeneous it will have been disturbed by generations of construction and alterations to the drainage.

Often the biggest obstacle to satisfactory remediation is the fact that a site is but one small piece in the urban mosaic. The substance which has been allowed to contaminate a site will probably migrate to neighbouring sites, and at some point will reach surface water. The groundwater and the surface water may be used for drinking purposes. The irony is that attempts to clean up a site may mobilise a contaminant that otherwise posed no immediate health risk.

We now have a much broader array of techniques available for treating contaminated land without the risk of removal to a special treatment plant or simply relocating the material to a contained facility. Although time-consuming and expensive many of these techniques are no more expensive than the old, unsatisfactory methods of relocation, or incineration. The other important lesson is that prevention is much cheaper than cure.

5.2.3 Liability

Remediation initiatives will succeed only if the regional context and the liability issues are addressed and this can happen only if the stakeholders agree to make it a priority. This means that there should be agreement on a regional scale that local taxes will not distort the costs to make greenfields cheaper to develop, as is the case round Toronto (De Sousa 2000). The tax situation is no better in the UK where VAT is charged on materials used in the restoration of a brownfield site, but not on the purchase of a greenfield site. Second, once the site has been remediated to the required standard the developers should receive an agreement from the authorities that they will not be liable for further work due to an upward re-evaluation of those standards. There is always some uncertainty as to how much it will cost to clean up a site because the full extent of the contamination may not be clear until excavation begins. That uncertainty can now be insured against by cost-overrun policies that have been developed by several insurance companies.

Despite these complications, physical, mechanical and biological technologies exist that can remediate most contaminated sites. It may be a lengthy and expensive process but the work can now be done *in situ*. The days of trucking all the material away – while dispersing some of it on the wind – and burying it elsewhere, are coming to an end. The deeper problem lies in agreeing on responsibility for any issues *in the future*, after a clean-up operation has been carried out in compliance with the local regulations. Some authorities issue a certificate of compliance that limits any future liability of the owner and the developer. Other authorities have refused to do this. In some places the notion of 'relative risk' has provided common ground for various parties in the government, the private sector and the local community. This approach recognises that an individual is surrounded by a multitude of risk factors, many of which will greatly exceed the threat of contaminants from a reused urban site after it has been cleaned up.

There still remains a huge problem, worldwide, in bringing wasted urban land back into use, thereby reducing the development pressure on the countryside. In general there has been more progress in Western Europe than in North America, as pressure on land use is greater in Europe, and people are more willing to live at higher densities. Within North America, Canada lags behind the United States in developing local partnerships to agree on a remedial course of action (De Sousa 2000).

5.3 Landfills – yesterday's solution

Ancient cities of the Middle East grew steadily higher as successive buildings rose on the discarded waste of earlier generations, forming 'tels' that provide archives of ancient life for today's archaeologists. In settlements that have not paved over the land surface for ease of walking and driving, solid waste – what little there might be – can be safely buried in the ground right by each house. As long as no contaminants are present most solid waste can be left to decompose. In Dakar, Senegal, they excavated the city's first landfill and reused it for compost. Only as cities become very large, their surfaces paved over, and their waste streams contaminated, does solid waste become a large and growing problem. This is exactly what happened to the modern industrial city.

However, a solution appeared to be at hand, or, at least, not too far away. The modern twentieth century city dealt with the solid waste problem by relocating the wastes to the rural periphery to rot in what Americans call landfills, and the British call (more accurately) rubbish dumps. The solution was quite cheap, such that, even today, few urban households know how much they are paying for this service, or where their wastes are taken. Like water supply, waste collection was a public service inherited from the days when costs were lower and the Public Health Idea (i.e. *urban* public health) determined priorities. Solid waste relocation becomes a problem as cities grow outwards to surround these once distant dumps. As society becomes richer, people generate more waste. As noted in Chapter 2, the amount of solid waste generated per household is a linear function of income. In addition to household waste there is commercial and institutional waste, industrial waste, and special wastes from hospitals and nuclear facilities. The last two are treated separately, but for many years all the other wastes might go to the same site, essentially just a big hole in the ground.

This heedless practice became problematic for two reasons. First, the communities that were chosen as unwilling hosts to the dumps began to object when new dumps or extensions were proposed. This was the origin of the not-in-my-backyard syndrome (NIMBY) in the 1970s. New locations, even extensions, became very difficult to obtain, especially in the suburbanising periphery of large cities and city regions. Residents objected to the disruption of the dumper truck traffic, the odour, the scavenger birds, and the impact of the perceived stigma on the resale value of their houses. The first factor – widespread discontent from the host communities – was the background to the second, fiercer objection to dumps, that being fear of the possible health impacts of contaminants, such as potential carcinogens.

The later factor was driven not by the countless municipal dumps across the landscape but the notorious case of the Love Canal disposal site in 1978, on the American side of the Niagara Peninsula. This was no ordinary landfill.

> The principal company that dumped wastes in the canal was Hooker Chemical Corporation, a subsidiary of Occidental Petroleum. The City of Niagara and the United States Army used the site as well, with the city dumping garbage and the Army possibly dumping parts of the Manhattan Project and other chemical warfare material. In 1953, after filling the canal and covering it with dirt, Hooker sold the land to the Board of Education for one dollar. Hooker included in the deed transfer a 'warning' of the chemical wastes buried on the property and a disclaimer absolving Hooker of any responsibility. . . . The elementary school was located near the center of the landfill. (CHEJ 2000)

Later:

> wastes . . . leached out into the basements of houses built later on the site, not only causing neurological and reproductive problems, but also being implicated in carcinogenesis and damage to the kidneys and urinary tract. (Connell *et al.* 1999: 9)

The presence of a primary school on the site accentuated the concerns of the residents. Eventually the school was closed and the federal government bought out 900 families. Occidental Petroleum has paid about $250 million in compensation and clean-up costs and also assumed responsibility for a remedial treatment plant on the site.

The upshot of these forces – the background of discontent, and the acute fear of harm to the children – changed the nature of the solid waste business quite abruptly, especially in North America. It became more widely known that the traditional dump was even more of a problem than nervous residents might suspect. As solid waste decomposes, it emits gases such as methane, which are potentially explosive. From the base of the dump, liquids, under pressure, flow outwards, as leachate, creating the potential for contaminating water supplies. Neither the gas nor the liquid fractions of the waste need be problematic, but the potential is there. The pressure was on for engineers to design 'state-of-the-art' landfills where these problems would be anticipated and eradicated. Methane could be trapped and burned to provide electricity (see Chapter 8). Similarly the floor and sides of the landfill could be lined with clay or other impermeable material, and the leachate channelled away and disposed of safely. Money could be spent on landscaping so that you would hardly know it was there. At least one local authority in Ontario paid to train hawks to deter scavenger birds from a landfill. Richer compensation was offered to host communities such as community halls and recreation facilities, as well as a share of the revenues. These modern landfills also created valuable jobs, at the manager level. All of these improvements meant that the cost of dumping solid wastes went up considerably. Special wastes are no longer acceptable at municipal dumps and so they have to be treated at specialised facilities at much higher cost, often 10 times the standard fee, and sometimes as much as 100 times more.

Unfortunately all these 'improvements' failed to solve the root problem. They are, after all, end-of-pipe solutions. We should not be creating, transporting and burying these materials in the first place. Nor do we have to. The answer, as indicated in Chapter 2, is to reduce, reuse and recycle, *not just to relocate*. The future lies in not creating the waste in the first place.

There is another problem, similar to the problem of contaminated urban land. That problem is the historical legacy of countless dumps dating back to the beginning of the industrial revolution. Many of these are unrecorded. The legal fallout from the Love Canal case stimulated surveys throughout the industrial world to attempt to determine where any such 'toxic time bombs' might lie (Swaigen 1995). In the United States over 14000 hazardous waste sites were mapped and the Superfund was set up to support remediation. Just as in the contaminated urban land situation, the question of responsibility soon overshadowed the technical questions concerning treatment. Estimates for Superfund expenditures suggest that 85% of disbursements, to date, have been for legal and administrative fees:

> Superfund spent $13.6 billion to clean up 155 sites out of their 1200 worst. It's estimated that 85% of this money was spent on fees, and very little of it actually went into the ground. It's estimated that the priority list for EPA is going to cost $64 billion, exclusive of military installations. (Robert Patzelt, quoted in Insurance Bureau of Canada 1995: 14)

Although current waste management standards are much higher than in the past there is little evidence that Western society as a whole is ready to deal with the problem at source by adopting a life cycle approach to all the materials we use. This approach implies that the producers (manufacturers, shippers and retailers) use only materials that the consumers can reuse, recycle or return to the producer.

5.4 Problematic building materials

5.4.1 Asbestos

The theme is a recurrent one. What seemed like a really good idea at the time turns out later to have hidden costs that eventually overwhelm the initial apparent value of the innovation. Nowhere is this unfolding more starkly apparent than in the discovery and use of asbestos, a naturally occurring, ductile mineral that was widely used in construction, transportation and manufacturing from the 1920s onward (Case 1999). Although not strong enough to be load-bearing it was used in combination with other materials to insulate against electric shock and as protection against fire. There is no doubt that its use saved countless lives from risks such as fire at sea. Unfortunately the fine fibres could also be inhaled. Once in the lungs they could lead to cancer and to asbestosis, a fatal inflammation of the lungs.

However, the period of latency approximated thirty years, so the first mortalities and autopsies – for asbestos miners – did not happen until the 1950s. Some of the victims were smokers and thus the cause of death might have been unclear, but gradually the evidence piled up. The circle of affliction spread to workers in the construction, transportation and manufacturing industries. Liabilities reached billions of dollars, much of it paid through workers' compensation. Risk to the general populace was still deemed to be low. The first case for outdoor or 'environmental' exposure was settled in the UK in 1995 (Aldred 1995). By then asbestos was being removed from buildings, using specially equipped, trained and certified firms. For the insurance industry this was their first significant – that is potentially destabilising – environmental exposure. Awareness of the problem is now widespread and therefore preventative measures for asbestos removal are required in most Western countries (DETR 2000).

5.4.2 Lead paint

Asbestos losses have not yet peaked but a second major problematic building material is already credited with causing major damage to human health, namely lead paint. Lead has been used as a paint additive since the 1880s because it enhances lustre, protects against mildew, and improves adherence to the surface material. There can be few buildings put up in Western cities between 1900 and 1970 that do not contain lead paint. As long as the paint remains attached to the surface to which it has been applied then it should cause no problem. However, as the paint becomes chipped, or the material to which it adheres itself deteriorates, flecks of paint are mobilised. The smallest may be inhaled, other bits can be picked up by inquisitive children and eaten.

Once this happens the paint is carried to blood, bone and soft tissue in the kidneys, liver and brain (Chia 1997). In sufficient concentrations it can impair children's mental development permanently. Recent studies have reported an association between lead levels in blood and attention deficit disorders, anti-social behaviour and crime. Many of the children who first appeared to be suffering from lead paint poisoning came from poorer families who were living in poorly maintained buildings. Like the miners with asbestosis, their symptoms could be associated with other potential aetiologies. In this case the host of deficiencies associated with poverty, such as poor nutrition, could all impede neurological development. Only a blood test made the diagnosis definitive. Unfortunately by then damage was generally permanent, although ameliorative treatments, such as chelation therapy, are available. In this case the financial costs have far exceeded asbestos for the cases that have been settled, because the most sensitive targets are very young children. Hence, the costs to the family include lifetime care and suffering.

Lead has been widely used throughout human history, beginning with lead pipes to carry water. However, the amount of lead that is bioavailable to humans has risen in modern society – even though we no longer lay lead water pipes – because we use it as an additive in gasoline and in paint. It is reported that: 'our bones today contain, on average, 30 times more lead than those of Egyptian mummies' (Lecomte 1999: 4).

As far as prevention of further cases is concerned, the response is the same as for asbestos. The material must be removed from the building using certified companies. The use of lead paint was banned in most Western countries in the 1970s. However, the typical rate of turnover of a country's building stock is about 2% per year, so it could be up to 100 years before this problem can be put behind us – 50 years for housing stock turnover, plus 50 years for the life expectancy of the last victims.

5.4.3 Other indoor problems

Asbestos and lead are only the two biggest materials errors to be introduced into our building stock. Other potential problems include electromagnetic fields, visual display units, PCBs, the multiple suspects for 'sick building syndrome', and others. Some are admittedly very difficult to link to verifiable health impacts. But that is not the whole story. The common response from industries that are accused of having introduced a damaging or fatal substance into the environment is *to deny that any effect exists*. This denial can go on for decades as the evidence mounts up. Our modern society is conditioned by economics to think short term, while – given the scientific uncertainties – a defence lawyer is likely to counsel his or her client to 'admit nothing'.

From an ecological perspective it would be better to regard every introduced material as suspicious until it has been proved innocent, rather than the other way round.

5.5 Underground storage tanks

Like other technological innovations, underground storage tanks seemed like an excellent idea at the time. They provided a safe storage environment for oil, which is flammable and potentially explosive. For petrol stations the safest thing to do seemed to be to set the tanks underground, far away from accidental ignition by a starting engine or a careless cigarette. During the last century they spread throughout the Western world, both for commercial operations like petrol stations, and also for homeowners who used oil for space heating, as was the norm throughout North America until gas began to be distributed by pipeline in the 1960s.

Unfortunately there was no way of checking the integrity of the tank once it was buried. Over time, ageing iron and steel tanks have corroded and some have failed, releasing their contents to the surrounding soil and rock fissures. Pollution of watercourses began to be traced back to particular tanks. Some leaks were discovered because of otherwise unaccountable loss of contents.

As such the tanks became a physical problem, especially on abandoned urban land where it could be difficult to find a solvent and available owner to take responsibility. Vacant land could be sold without any indication of tanks left in the ground. Even if owners could be found and those owners had insurance policies to cover themselves for damage to third parties, the problems did not end there. Like the damage associated with leakage from landfills, responsibility for this particular type of environmental problem hinged on a crucial point of insurance law. Insurance policies are designed to cover unforeseeable events, like genuine accidents that occur suddenly. They specifically exclude damages arising from 'wear and tear' or negligence on the part of the insured party.

Underground tanks have proved to be a thorny problem for insurers because insured parties could certainly claim ignorance of what had happened underground to their tanks. They had not neglected their tank, so much as never thought about it. The insurer could counterclaim that leakage was something that happened gradually over the years and therefore did not fit into the 'sudden and accidental' event for which coverage had been provided. Some jurisdictions have found for one party, some for another. From the urban infrastructure point of view, it shows the need to constantly modify how our systems are operating, even if they are 'underground and out of mind'.

Like the asbestos and lead problems, a long term solution lies in demanding higher standards from now on. In this case the technical solution is not complicated. New underground storage tanks are provided with a double hull and a monitor that warns the owner if either hull is breached. The pressure between the walls goes down and remedial action must be taken. Insurers now require that new tanks operate under this system; otherwise coverage is not available.

5.6 Subsidence

Subsidence may occur in quite a variety of circumstances including natural erosion of the limestone bedrock and collapse of mining works such as coal in northern England and gold in Johannesburg. In the industrial age the over-pumping of groundwater may eventually cause considerable subsidence as can be seen in cities as far apart as Mexico City, Venice and Bangkok. A new agent of subsidence occurred in Britain in the 1970s due to the shrinkage of clay soil during unusually warm and dry summers. In retrospect this may be seen as one of the first visible aspects of global warming in Britain. The UK Climate Impacts Programme expects that, 'By the 2080s, summer potential evapotranspiration over southern England is 10 to 20 per cent higher than at present' (Hulme and Jenkins 1998: v). Unless this increase is accompanied by more rainfall it should be expected that clay shrinkage and subsequent subsidence will become a greater problem.

This problem serves to remind us that in addition to the past mistakes of the industrial revolution – such as contamination – we are now seeing the impacts of a deeper and more far-reaching problem in climate change. Until recently Britain could be regarded as a cool and moist country, not given to the climatic extremes of the tropics or regions of Mediterranean climate. Construction went ahead on the assumption that conditions would remain cool and moist; now we can no longer assume this.

Again, the insurance industry has been the first sector in the economic system to have to deal with this new problem. In the UK insurers have paid for expensive underpinning of damaged properties and have tried to identify the builders whose structures have failed – a fairly hopeless task for houses built fifty or more years ago. Apart from the physical consequences of the damage, there is also the fear of resale values being affected by the stigma, similar to the problem associated with the proximity of housing to landfills, mentioned above. A recent study suggests that British householders in the clay belt might have to get used to cracks in their houses and just patch them up at their own expense, as is the practice in countries like Australia, South Africa and the United States (Association of British Insurers 2000).

5.7 Conclusion

The technology associated with the industrial revolution has introduced a large set of new materials into the environment while significantly amplifying the flows of naturally occurring substances. Our knowledge of the movements of many of these materials and their potential impact on human health and ecosystem health is incomplete. Even where the key physical aspects of problems such as contaminated land and underground storage tanks are well understood, the solution soon becomes mired in the issue of responsibility. This reluctance to take responsibility stems from a situation where the original expectations for profit are thwarted by unexpected side effects, such as the toxicity of asbestos and lead paint. The typical initial response of the parties to the process is to deny any knowledge of adverse implications associated with their product.

The combination of these events lengthens the time it takes from initial suspicion of a problem to the development of an adequate protocol for minimising future effects and to recompense for those who have sustained any damage. However, it is becoming more evident that we must adopt a precautionary approach to all activities that potentially put people at risk. The first step in this direction is to develop a life cycle approach to the materials we handle in the urban environment. The question that arises is: what kind of life cycle can we envisage for urban land – inevitable decline and abandonment, or revitalisation?

5.8 Websites

1. Association of British Insurers (ABI): *http://www.abi.org.uk*

2. Canada: National Roundtable on the Environment and the Economy: *http://www.nrtee-trnee.ca/eng/home_e.htm*

3. Love Canal disposal site, New York State, USA: *http://www.essential.org/cchw/lovcanal/lcindex.html*

4. UK Department of the Environment, Transport and the Regions (DETR): *http://www.environment.detr.gov.uk*
   DETR Asbestos and man-made fibres in buildings: *http://www.environment.detr.gov.uk/asbestos/guide/idex.htm*
   DETR Waste strategy: *http://www.environment.detr.gov.uk/wastestrategy/index.htm*
   DETR Contaminated land: *http://www.environment.detr.gov.uk/landliability/index.htm*

5. USA Federal Government Remediation Technologies Roundtable: *http://www.frtr.gov/*

5.9 Further reading

Association of British Insurers (2000). *Subsidence: A Global Perspective*. General Insurance: Research Report No 1. London, ABI.

Case, P (1999). *Environmental Risk Management and Corporate Lending: A Global Perspective*. Cambridge, Woodhead Publishing.

Connell, D, Lam, P, Richardson, B and Wu, R (1999). *Introduction to Ecotoxicology*. Oxford, Blackwell Science.

Davis, M L and Cornwell, D A (1998). *Introduction to Environmental Engineering* (third edition). Boston, McGraw-Hill.

Kibert, C J (1999). *Reshaping the Built Environment: Ecology, Ethics and Economics*. Washington DC, Island Press.

Lecomte, P (1999). *Polluted Sites: Remediation of Soils and Groundwater*. Rotterdam, A A Balkema.

6

The air we breathe and the climate we are changing

For most children in the large cities of developing countries, breathing the air may be as harmful as smoking two packs of cigarettes a day. . . . A 1990 study of atmospheric lead pollution in Bangkok estimated that 30 000 to 70 000 children risked losing some 4 or more IQ points because of high lead levels. (World Bank 2000: 141)

Climate change is a global issue caused by local activities. The extreme weather conditions over the years can no longer be ignored and municipalities, including ours, need to take responsibility for the causes. Flooding, extreme snow fall, ice storms, extreme heat are events that seem to be occurring more and more frequently putting a strain on municipal budgets, staffing and emergency resources. I believe that this is just a beginning. From what I know of climate change we should all be taking the precautionary principle very seriously. (City Planner from Airdrie, Alberta, quoted in Robinson 2000: 132)

6.1 The issues and the impacts

The impact of our emissions to the air used to be a local matter. As the economy has become industrialised these impacts have gradually spread from the city itself, to the surrounding regions, and finally to affect the global circulation of the atmosphere. We are now responsible for multiple impacts at all spatial and temporal scales from the diurnal and local to the irreversible and global. These various emissions interact with one another. Some gases have conflicting effects, in the sense that some cool the atmosphere while others warm it. The warming effect itself will have adverse consequences for air quality. Most of these gases are the result of fossil fuel combustion and associated industrial processes such as smelting and cement manufacture.

Discharges of particulates to the air have long been associated with excess deaths from respiratory and cardiac stress. The worst excesses of the 1950s led to higher air quality standards in the West, but conditions continued to be very poor in developing countries and in communist countries. Although the toxicological and epidemiological evidence is often disputed in the West, the situation revealed by the collapse of communism in Europe is incontrovertible, as illustrated by this report on the Voivodeship (province) of Katowice, Poland:

> The 1989 infant mortality rate was 25.5 per 1000 live births in
> Katowice City, with an average of 18.5 in the Voivodeship; this
> compares unfavorably with the national average of 16.1 per 1000. The
> results of a six-year study in the 1980s in sub-regions of the Voivodeship
> conclude that infant mortality is correlated with dust fall, ambient
> levels of lead, tar, phenols, formaldehyde and benzo(a)pyrene . . . The
> Voivodeship also suffers from the highest incidence of premature births
> (8.5%), genetic birth defects (10.1% of all live births), and spontaneous
> miscarriages in Poland. (Borkiewicz et al. 1991: 19)

At one time any concern about these impacts could be shrugged off as being an unavoidable price to pay for material 'progress' – unwelcome, but not important in the great scheme of things. In institutions like the World Bank in the 1970s you could not raise an environmental issue like air quality without being accused of disloyalty to the Development Project, the mission of the Bank. The general assumption in such organisations was that it was all right for rich countries like Britain and the United States to set standards for air quality, but it was an inappropriate luxury for the developing world (*The Economist* 1992a, 1992b). Now the Bank admits that air quality in some cities could have the equivalent impact on children of smoking two packs of cigarettes a day (see the introductory quotation to this chapter). From Bangkok to Mexico City, lead in the atmosphere could affect the mental health of 'tomorrow's human resources', otherwise known as children.

Furthermore the Bank concedes that climate change could also be a serious problem:

> Climate change is occurring at unprecedented rates because huge
> quantities of carbon dioxide, methane, and other greenhouse gases are
> being released into the atmosphere daily. Global temperatures have
> been rising slowly since 1800. The 20th century has been the warmest
> century in the past 600 years, and 14 of the warmest years since the 1860s
> occurred in the 1980s and 1990s. Temperatures in 1998 were higher than
> the mean temperatures for the 118 years on record, even after the effects
> of El Niño are filtered out. Satellite readings now confirm a similar
> elevation of temperatures in the upper atmosphere. . . . The source of
> the increase in carbon dioxide thus far . . . is anthropogenic. These facts
> are now widely accepted. (World Bank 2000: 41)

This shows that perceptions of the problems caused by emitting waste gases to the atmosphere have changed enormously over the last twenty-five years. If a conservative institution like the World Bank admits that we have a problem, you can be absolutely sure that we do. It means that the expected impacts will be considerable.

None of this should come as a surprise because we know that our industrial technology takes us through a predictable pollution transition as various stages of industrialisation are reached. The burning of wood and coal as household and industrial fuels produces particulates and sulphur. Eventually cleaner fuels and scrubbers reduce the output of these residuals. In the next stage the major problem source is the burning of petroleum in the internal combustion engine. This produces hydrocarbons, nitrous oxides, lead, sulphur, carbon monoxide and carbon dioxide as residuals. Some of these can be turned into less harmful residuals by the use of catalytic converters. The one residual that has so far eluded capture is carbon dioxide, the release of which has risen inexorably as income rises. In Canada, this collection of impacts has been summarised as the 'six air issues', comprising smog, suspended particulate matter, hazardous air pollutants, acidic deposition, stratospheric ozone depletion and climate change (Munn *et al.* 1997; IES and Environment Canada 1999; see Fig. 6.1).

The first half of this chapter will focus on the local problems that directly affect human health. The second half will deal with the regional and global impacts of our atmospheric emissions.

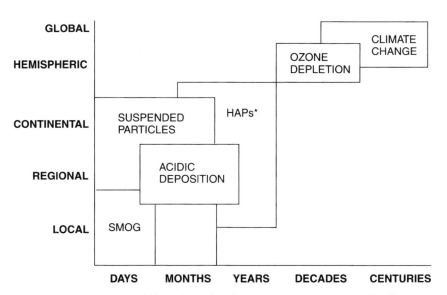

* Hazardous air pollutants, such as mercury and lead

6.1 The 'six air issues'

Source: after A Maarouf in IES and Environment Canada 1999.

6.2 The mounting cost of poor health

The first large scale health impacts recorded in the post-war period were the urban smogs (smoke + fog), formed in moist conditions around the particulates released from burning coal. Major smogs in the late 1940s and early 1950s led to the establishment of air quality standards and the banning of coal as a domestic fuel (see Table 6.1). Later, people became more cynical about these excess deaths as they mostly happened to the elderly and other susceptible population groups who would have died soon anyway, and thus the polluted air had simply 'harvested' them early. However, in less cynical times it came as a shock to realise that smog could kill.

Other impacts took longer to understand, even in places like Los Angeles where an enclosed site, warm weather and the highest per capita car ownership in the world created the perfect laboratory for making ozone from the effect of sunlight on the hydrocarbons emitted from vehicles. Although it was not disputed that elevated ozone causes irritation of lung tissue it was some years before the mortality impact was established and then it was found that the 'susceptibles' included children and even young adults, people who were in the prime of their life. Later came the even more astonishing discovery that the impacts extended beyond the city into rural areas (City of Toronto 1998; IES and Pollution Probe 1998).

In the UK, as we saw in Chapter 3 (Table 3.4), air pollution is the leading external cost imposed by road transport, accounting for nearly half of the total cost. The derivation of that sum is broken down in Table 6.2, which details the costs of the principal air pollutants in terms of morbidity and mortality.

Mexico City provided the first developing country example of an urban system that was suffering seriously from inferior air quality. Like Los Angeles it was sited in a warm, enclosed space, but what made it even worse was the high elevation. The air became so bad that oxygen kiosks were installed in the downtown area. Lead from automobile exhaust was established as the culprit for reduced IQ among school children (Guerra 1991).

Table 6.1 Major air quality incidents and the introduction of Clean Air Acts

Date	Event
1930	Meuse Valley, Belgium. Smog. 63 deaths
1939	London, England. Smog. 1300 excess deaths
1948	Donora, Pennsylvania. Smog. 20 deaths
1952	London, England. Smog. 4000 excess deaths
1956	UK Clean Air Act passed
1963	USA Clean Air Act passed
1971	Canadian Clean Air Act passed

Source: Henry and Heinke 1989: 473–4.

Table 6.2 Estimated external health costs per pollutant for the UK (all figures are in 1993 prices)

Pollutant	Effect	Number of premature mortalities per year	Total external health costs £ millions
Direct PM$_{10}$	Mt	1725	3450
	Mb		2100
SOx (inc ind PM$_{10}$)	Mt	1880	3760
NOx (inc ind PM$_{10}$ and ozone)	Mt	2000	3990
	Mb		2160
VOCs	Mt	1010	2020
	Mb		850
Lead	Mt	20	40
	Mb		240
Benzene	Mt	30	70
Total		6665	18680

Source: Adapted from Maddison *et al.* 1996: 75, based on Calthrop 1995.

Mt = mortality, Mb = morbidity.

Although Mexico City became the first city in a developing country to become notorious for air pollution, it was not alone. Collaboration between the World Meteorological Organisation and the World Health Organisation provided evidence of serious impacts in 30 major cities. The World Bank followed this by creating a new environmental indicator, the DALY (for 'disability-adjusted life years'), which added the days lost through disability to reduced life expectancy in poor countries compared with low mortality rich countries (World Bank 1993). Of the burden of disease attributable to environmental conditions (including occupational and household settings), the Bank found that 36% could be attributed to poor urban air quality. In total about 1.3 billion urban residents worldwide are exposed to air pollution levels above recommended limits (World Bank 1993).

Further shocking evidence of the impacts of poor air quality came from Central and Eastern Europe after the collapse of communism in 1989, as quoted above. A lot of the pollution there was attributable to uncontrolled emissions from industry. Action was taken to reduce this and the implosion of the former communist economies speeded the process considerably. Unfortunately, once the economy began to restructure and become market-oriented, car ownership and use soared, thereby exchanging one set of pollutants for another.

From all of these urban experiences with air quality we can say that we have the technological means to control the emission of residuals like lead, sulphur and the larger particulates. What we have been unable to do is curb the vehicular emissions like nitrous oxides, other ozone precursors, carbon dioxide, and particles less than 10 microns, which are now thought to be responsible for lung impairment. Nor are we travelling in the right direction to do so. Although higher prices and more stringent regulation in some Western countries have encouraged the development of more fuel-efficient engines, car ownership is still growing and kilometres per year per vehicle are increasing. In poorer countries the age of mass ownership of private automobiles and the development of long distance road freight transport are only just beginning.

Even though some air quality indicators have improved in richer countries, emissions from power stations and motor vehicles pose a grave threat especially in larger cities that are still growing and where automobile dependency remains high. For example, Toronto is mainly a twentieth century city, owing comparatively little to the nineteenth century for its industrial base. Hence its air quality problems are largely attributable to motor vehicle dependence, rather than heavy industry. Even this was considered to be a manageable problem until hospital records indicated otherwise. As Jeffrey Brook reported:

> There is a history of at least 15 years of health effects research based upon over 20 years of data in the Toronto–Niagara Region. This research has clearly demonstrated that a number of different air pollutants, at concentrations observed in the region, are associated with respiratory and cardiac health effects. (Brook 1999: 81)

The relevant pollutants identified by this research included ozone, sulphur dioxide, particulates, particle acidity, carbon monoxide and nitrogen dioxide (Brook 1999: 83).

The hospital admissions data are known to be part of a 'pyramid of health effects', fewer in number than the actual number of sufferers on a given day, but many more than the number of deaths attributable to poor air quality. It has been estimated that the mortalities attributable to poor air quality amount to 800 people per year in Ontario.

6.3 Air quality management

The management of air quality – urban and regional – presents fundamental policy challenges for all levels of government. The more far-reaching of these challenges – including a different approach to urban land use – will be analysed in Chapter 8. At this point we are assessing the options for dealing with the problems of the past, specifically the political and technological choices that have brought us to the present situation. Clearly urban governments control a number of policy levers relevant to air quality, including land use and transportation planning. Many Western cities have the authority to shut down polluting factories on bad air quality days; and some can even forbid the use of private transportation. These are short term decisions. In the medium term it is usually the responsibility of the local authority to provide attractive – or at least adequate – public transport, either through a public authority or by licensing private carriers. The steady rise in car ownership and car use in the West suggests that most governments have failed to do this. Whereas there may be a few automobile-lovers who are quite happy to sit in traffic jams, it would seem safe to assume that most would take public transport if they felt that it met their needs (such as convenience, safety and comfort), at a reasonable price. In many North American cities public transport is still regarded as the refuge of the car-less poor and elderly.

Some urban governments are attracted to congestion charges because they hope that they will provide revenue to fund improvements in public transport while discouraging driving in the central part of the city. The problem with this approach is in the timing – the charges may be levied for years before the improvements take place, if they do take place at all. What is needed is a business plan that would secure the funds to begin improvements at once, using a loan against future revenue from fares and congestion charges.

There is another factor, apart from air quality, that requires changes in the system. Motorised individual transport does not provide an adequate means of mass transportation anyway. The technology – the vehicles themselves, roads, parking areas, petrol stations – requires far too much space to provide adequate circulation throughout the urban system – just as described by James Lovelock in his portrayal of all land-based organisms (introductory quotation to Chapter 2). With 80% of private vehicles carrying only the driver there are just too many vehicles trying to crawl through the same space. As a luxury conveyance 100 years ago, the automobile brought much pleasure to the very few who could afford it; but once automobiles are affordable for the mass of the population most urban systems seize up. There is no mystery in this. There is no solution in building more roads or installing more traffic lights. It is true that you can build urban freeways, as has been done widely in the United States. The problem with this 'solution' is that it destroys the city and replaces it with a collection of highways and parking lots surrounded by office blocks. In Toronto about 40% of the land is devoted to automobile infrastructure, compared with only 7% to open space and parks (City of Toronto 1998).

Many of the options to change this situation lie with local governments. However, other complementary policy levers are often in the hands of higher levels of government – regional/provincial, national and international. These policy levers include energy taxes (including petrol), air quality standards, regional transportation planning, and the privatisation and regulation of energy utilities. Any one of these levers could be pulled in a direction that might negate local governments' attempts to improve air quality. For example, air quality in the Toronto–Niagara region will be affected by the deregulation of Ontario's electricity supply system that will pass from a crown corporation (a parastatal) to the private sector (IES and Pollution Probe 1998). This will require vigilant supervision of emissions and resultant air quality if even the present, inadequate standards are to be maintained. In the Ontario case international relations are an important factor because the relevant air-shed includes the Ohio Valley in the United States, which is upwind of southern Ontario, the home of most of Ontario's urban population. Ironically, it is possible that a shortage of generating capacity in Ontario will lead to increased imports from the Ohio Valley – much of it from older generating stations – which will further reduce air quality, because old stations do not have to meet current emission standards.

6.4 Regional and stratospheric impacts

Of the six air issues identified at the beginning of this chapter, it could be argued that three of them are not controllable by urban governments, those being acid deposition, stratospheric ozone depletion, and the long range transport of hazardous air pollutants. To the extent that this is true they will not be discussed in detail. However, all are connected with modern industrial life and are therefore linked to many of the issues that are specifically a function of modern urban living. For example, the amount of energy that urban users demand is an important fraction of total energy demand. The heating and cooling requirements of our buildings and appliances directly affect the quantity of fossil fuels burned and the CFCs, and CFC substitutes, used. CFC substitutes contribute to the enhanced greenhouse effect.

Acid particles like sulphates and nitrates have a corrosive effect on buildings and an unknown effect on human beings. From the 1960s onwards it became clear that they also had damaging impacts, directly on vegetation and indirectly on fish. Acidified forests exhibited regional dieback and lakes became lifeless. Legislation set caps on emissions and gradually the worst local impacts were eliminated in the wealthier countries by the introduction of scrubbers on smokestacks. However, these emissions are still a problem in cities in developing countries and they continue to have major regional impacts in richer countries too.

The principal source of sulphur dioxide is power production from fossil fuels (70% of the source in the United States), while the main contributor to nitrogen oxides (43%) is transportation. Predictably, more progress has been made on SO_2 than NOx, because it is easier to monitor emissions from stationary sources (power plants) than mobile sources (automobiles and trucks), and also the regulation of power plants affects the voting public less directly than the use of private vehicles.

It is generally felt that governments – both national and international – have had some success in responding to acid deposition and stratospheric ozone depletion. Certainly there are lessons to be learned from these experiences that could be useful in controlling greenhouse gases. For example, private corporations co-operated with legislative bodies to design a solution to acid deposition, and that solution included market mechanisms to reduce emissions in an efficient manner, using the 'cap and trade' approach. However, as noted, what worked with corporations did not work so well with the motorised public. This is a problem that no government has yet confronted, except a few municipalities, like Curitiba, and the city-state of Singapore (Rabinovitch 1992; Rabinovitch and Hoehn 1995; Rabinovitch and Leitman 1996).

In two other respects these issues affect the management of greenhouse gases. First, sulphates have a cooling effect on atmosphere, and thus their removal is likely to enhance global warming. Second, the main substitutes for CFCs are hydrofluorocarbons (HFCs) which are powerful greenhouse gases (see Table 3.1).

6.5 The changing climate

The problems of acid deposition and stratospheric ozone depletion are serious and long term. Both could eventually result in major biospheric changes that could be very adverse for humankind. However, both are fairly well understood and manageable compared to human induced climate change. This is an altogether larger dislocation of the intricate relationships that support the biosphere. Fourier first predicted the enhanced greenhouse effect due to human burning of fossil fuels in 1826. The issue was raised intermittently in the intervening years but it was not until the early 1950s that scientists realised that we were emitting carbon dioxide faster than the ocean could absorb it from the atmosphere, and hence it must be building up in the atmosphere itself. The first direct measurements in the lower atmosphere to verify this hypothesis were begun in 1957, and we have been able to monitor the steady accumulation of carbon dioxide in the atmosphere in detail ever since.

It has taken nearly another fifty years for the global community to allow itself to be convinced by scientists that this is indeed a serious problem that will change the world as we know it, in ways that we cannot reverse. We are now committed to climate change from the long-lived trace gases that we have ejected into the atmosphere, and we are still increasing the rate at which our emissions continue.

The impacts of these emissions are intricately linked and far from fully understood. They will vary by region and over time. What we do know is that most places on earth will undergo climatic changes that will affect every aspect of people's lives, and in some cases will be life threatening. The best understood impact is that surface air temperatures and water temperatures will rise. In Britain, for example, the biggest initial increase will be seen in night-time winter minimum temperatures. The following is a typical regional scenario:

> The rise in annual mean temperature in the UK is slightly smaller than the global average, with a pattern of larger increases in the southeast of the country than in the northwest. Thus by the 2080s, annual temperatures over southeast England are between 1.5 °C and 3.2 °C warmer than the 1961–90 average, but over Scotland only 1.2 °C to 2.6 °C warmer. The warming is generally slightly more rapid in winter than in summer and greater during the night-time than day-time. The year-to-year variability of temperature increases in summer, but decreases in winter. (Hulme and Jenkins 1998: v)

In some parts of the world, such as the UK, the temperature increase may improve productivity for some crops, but that depends on water availability and on the impacts of climate change on pests. The temperature increase will also change the distribution of disease vectors that affect humans, such as malaria and Lyme disease.

The temperature rise will be higher in the continental interiors that include the world's major granaries. Forest fires will become more frequent while forests will be further stressed by pests. If there is a net loss of forest biomass then this will provide a positive feedback to the greenhouse effect through the additional release of carbon to the atmosphere.

The warmer temperatures will produce a rise in sea level due to thermal expansion of the oceans, plus the melting ice-caps and glaciers. It is now believed that the Arctic ice-cap has lost 40% of its mass during the last thirty years. It is also expected that a warmer world will produce more extreme weather events, such as hailstorms and tornadoes and possibly tropical cyclones. It will definitely produce more droughts, and they are by far the costliest of extreme weather events even though they receive less publicity than destructive storms. Overall, a warmer world will lead to an intensification of the hydrological cycle, which has profound implications for water management worldwide.

All of these predicted changes have immediate consequences for urban environmental management and the building of ecological cities. The changes will clearly make the task much more difficult. Warmer cities, subject to longer dry spells and more intense rainfall will require major modifications to their infrastructure and to the flows of people, water and materials. Those cities that will be directly affected by sea level rise will have the most adjustments to make because all their drainage systems are designed for current sea level.

The most alarming aspect of this impending change is the degree of unpredictability. Despite countless completed and ongoing regional studies we still know very little about what *will* actually happen, rather than what *may* happen (Maxwell *et al.* 1997; Watson *et al.* 1998; Parry *et al.* 1991). The uncertainty is partly due to uncertainties about the feedbacks between climate change and the biosphere, such as the release of methane from the permafrost of the peri-Arctic, or the release of methane hydrates from the ocean bed as it warms up (Houghton 1994: 35; Leggett 1999: 45). We do not even know if the release of additional freshwater from the melting Arctic ice-cap will deflect the Gulf Stream that keeps north-western Europe warm. However, the biggest uncertainty is humankind itself. At what point will people take the threats seriously enough to take appropriate action to curb emissions? The most critical decisions relate to our urban–industrial lifestyle.

The remaining sections of this chapter will look at several of these climate change issues in turn:

- living with higher temperatures in the city
- impacts of extreme weather events on cities
- opportunities for reducing greenhouse gas emissions from the city.

6.6 Living with higher temperatures

The change of which we are most certain is that temperatures will increase generally, and in the continental interiors of the northern hemisphere more than elsewhere. We have also created a local temperature increase in our cities through the heat island effect. For cities like Chicago or Moscow, far from the sea and located in the northern hemisphere, this spells hotter summers as well as the possibility of warmer winters.

Chicago experienced a preview of these conditions in July 1995 when temperatures for five days reached 34 °C or higher and the daily minimum did not fall below 24 °C (NOAA 1995). The problem was intensified when a power station shut down, overloaded by the cooling demand. Despite the fact that the city set up air conditioned tents, 465 people died due to heat stress and associated heart failure. The victims were mostly poor, elderly people who lived alone. Of those who died dozens were left unclaimed by friend or relative and were eventually buried in a common grave by the city authorities. Milwaukee recorded another 85 deaths from the same event, while the US total heat-related mortalities for the summer reached 1000, compared with the normal yearly total of 175.

Much was made of the fact that the EPA predicted the heat wave and the city had time to make preparations for relief. The critical system failure was attributed to the unwillingness of the victims to leave their apartments for fear of theft of their property or assaults on their persons. Authorities in other cities in a similar climatic situation had to ask themselves: 'What would have happened here?' We will probably soon find out because such weather events are expected to become more frequent.

This event is only one example – especially pertinent to urban management – of the impact of recent 'heat disasters' in the continental United States. Table 6.3 provides a summary of the major events in the 1980s and

Table 6.3 Heat disasters in the continental United States since 1980

Date	Event	Economic cost $ billions	Deaths
Summer 1998	Drought	6–9	200
July 1995	Midwest heat wave	—	550
Summer–Autumn 1994	Western fire season	1	undetermined
Autumn 1993	Southern California wildfires	1	4
Summer 1993	Drought	1	undetermined
October 1991	Oakland firestorm	2.5	25
Summer 1988	Drought	40	5 000–10 000
June–September 1980	Drought	20	10 000
Total (approx)		73	

Source: NOAA/USDA 2000.

1990s. Greater heat provides two immediate challenges to urban management – the threat to health from heat stress and poorer air quality, and the threat to health from diminished and poorer quality water supply. Longer term threats include shifts in disease vectors, such as may be signified by cases of West Nile fever in New York City in 1998 and 2000.

In addition to the impacts of heat stress in the city we can expect greater urban exposure to forest fires as has been experienced recently in southern California (1993), and northern Florida (May 1998). Throughout the United States there is a recognition of an enhanced risk of the 'wildland/urban fire hazard' (ISO 1997). The problem is worldwide. Cities in the northern Mediterranean are also at risk, while the international repercussions of uncontrolled fires was demonstrated in 1999 when widespread burning in Sumatra reduced visibility, closed airports and affected human health in Singapore and other Asian cities.

6.7 Extreme weather events

Drought and heat waves – even on the scale of the Chicago event – tend to generate less news than the sudden, disastrous impact of tornadoes, hailstorms and hurricanes, because they are less photogenic and less of a shock to the rest of society. What is most certain is that higher temperatures in the continental interiors will produce more frequent, and perhaps more intense, convective storms – including tornadoes, hailstorms and summer rainstorms (White and Etkin 1997). The risk from hurricanes to urban systems is increasing – whether or not the frequency or intensity of hurricanes increases with global warming. This is simply because most cities are still growing, becoming more complex, and becoming more dependent on automobiles and on electricity. A modern city ceases to function if electricity is unavailable, while private automobiles are unable to evacuate the population when threatened by a sudden emergency such as a hurricane.

It is widely accepted that hurricane Andrew was a 'near miss' as it came close to striking both Miami and New Orleans. New Orleans was threatened again in September 1998 by hurricane Georges which made landfall only 50 km to the east; while '1.5 million people along the Gulf Coast were evacuated' (Partner Re 1998). Vulnerability of this magnitude requires an entirely new approach to urban risk management which goes far beyond the traditional scope of emergency response.

It is the very unpredictability of living with climate change that requires this new level of commitment. It is very difficult to prepare for the unexpected, but that is what is needed as was clearly demonstrated by the 1998 ice storm which struck the Ottawa Valley and northern Maine, and paralysed the city of Montreal for several days (Higuchi *et al.* 2000; Lecomte *et al.* 1998; Kerry *et al.* 1998). Although ice storms are a frequent occurrence in this and similar climatic regions, no one predicted an event of this magnitude. Because the deposition of ice continued over several days the weight on the electricity lines and pylons exceeded all previous experience, resulting in the collapse of the pylons and the widespread loss of power.

6.2 Another near miss. Hurricane Georges (September 1998) was heading directly for New Orleans until it veered at the last minute, making landfall 60 km to the east. New Orleans survived a similar near miss from hurricane Andrew in 1992.

Source: Partner Re. 1998.

The bridges linking the island of Montreal to the shore were blocked by traffic and ice. Even if vehicles could have been brought in there was no reception area with the capacity to receive the trapped population. Only 28 deaths were attributed to the ice storm on the Canadian side of the border, but it could very easily have been more if a serious fire had broken out in Montreal. The event generated over 800 000 insurance claims, even more than hurricane Andrew. Even so – in terms of impact – it was another 'near miss'.

6.8 Reducing greenhouse gas emissions from the city

It is 10 years since the City of Toronto produced its first plan to reduce greenhouse gas emissions (City of Toronto 1991). Some progress has been made in that time and this progress was recognised at the Cities for Climate Protection World Summit, held at Nagoya, Japan, in parallel with the Kyoto Conference of the Parties on the Climate Change Convention in November 1997. At Kyoto the Canadian government made a commitment to reduce greenhouse gas emissions by 6% from the 1990 level. Although no individual cities were represented in the delegation it is obvious that most of the reduction efforts will have to come from urban areas, as that is where 80% of Canada's population resides. Accordingly, a research group in Toronto developed proposals as to how that 6% reduction could be made within the Toronto–Niagara region (Timmerman *et al.* 1999).

Table 6.4 sets out the sources of emissions in the region by sector in 1990 and Table 6.5 lists the solutions that were analysed. As for other prosperous industrial regions, it was assumed that the contribution of industry to emissions would fall, while the share of transportation would rise, both commercial and freight. Population was also assumed to continue to rise in the

Table 6.4 Toronto–Niagara region (TNR) greenhouse gas emission estimates (kilotonnes of CO_2 equivalent)

Source of GHG emissions	1990 TNR levels kt CO_2	% of total TNR emissions
Power generation	12 098	17
Industrial	19 980	28
Commercial/residential/administration	12 823	18
Personal mobile	15 688	22
Commercial mobile	5 472	7
Landfills	3 050	4
Other	3 291	4

Source: Timmerman *et al.* 1999: 140; based on Khan 1998.

Table 6.5 Toronto–Niagara region (TNR): strategies proposed to reduce greenhouse gas emissions

Source of GHG emissions	Strategies proposed
Power generation	1. Integrated gasification combined cycle technology 2. Tradable permits 3. Develop or import renewable energy and hydroelectric power
Industry	1. Efficiency increases 2. Remove CO_2 pre- and/or post-combustion 3. Tradable permits
Commercial	1. Improved design re energy demand 2. Retrofits
Residential	1. Economic incentives to insulate and improve energy efficiency
Administration	1. Retrofits
Personal mobile	1. Increase average vehicle occupancy through employer co-operation and priority lanes for carpooling 2. Shift fuel mix to gas, hydrogen and electricity 3. Traffic calming and land use changes 4. Variable road pricing 5. Inspection and maintenance programmes
Commercial mobile	1. Demand management through pricing 2. Inter-modal technology (road/rail integration)
Landfills	1. Gas capture

Source: Timmerman *et al.* 1999: 142–3; based on Khan 1998 and Kumar 1998.

region. The group found that, technically, it was not difficult to make the 6% cut, if various assumptions could be made for the regulatory regime and the behavioural response. But therein lies the problem. The evidence suggests that there is little public appetite to make the behavioural changes and little government appetite to demonstrate leadership on the issue. As long as this situation continues we will remain trapped in the problems of the past and we will continue to pay for our inaction.

6.9 Conclusion

Even policy-makers in formerly sceptical institutions now agree that both air quality and climate change are important issues and that action to reduce these changes should be undertaken. The fact that our fossil fuel derived emissions to the air cause such a variety of problems presents us with an opportunity to gain multiple benefits from emission reduction. The big problem is that our technological success in producing more energy-efficient processes (such as automobile engines) cannot outweigh the increased emissions that derive from the continuing growth of the population and the growth of the economy. The transportation sector appears to be particularly difficult to bring under control for political – not technical – reasons.

One additional problem may actually help to identify a solution. Road transportation itself becomes deadlocked in traffic jams because the very density of city centres makes it impossible to identify enough space to provide the desired level of access. The only 'solutions' to this problem are freeways at multiple grades, which then destroy the city as a place of human residence and interaction. The combination of more efficient engines, more vehicles and more roads running through the urban system is clearly *not* a solution to any of the major problems – congestion, air quality or climate change.

Although it is relatively easy to 'make the emissions cuts' required by the Kyoto Protocol, this is not enough. First, any such reductions will rapidly be wiped out by continuing population growth and economic growth, if we continue with broadly the same technology. Second, the Kyoto round of cuts are just the beginning – the first step along the road to a society that no longer depends on burning fossil fuels. In order to travel along this road much more radical solutions will be required.

6.10 Websites

1. Canada's National Climate Change Process:
   http://www.nccp.ca/html/index.htm

2. Climate Ark – news service on climate change:
   http://www.climateark.org/

3. ICLEI Cities for Climate Protection Campaign:
   http://www.iclei.org/projserv.htm#ccp

4. NOAA, hurricane watch: *http://www.nhc.noaa.gov/*

5. Toronto–Niagara region study on atmospheric change:
   http://www1.tor.ec.gc.ca/earg/tnrs.htm

6. UK Climate Impacts Programme: *http://www.ukcip.org.uk/ukcip.html*

7. US EPA – Acid Rain Program:
   http://www.epa.gov/acidrain/overview.html

6.11 Further reading

Hulme, M and Jenkins, G J (1998). *Climate Change Scenarios for the United Kingdom: Scientific Report*. UK Climate Impacts Programme, Technical Report No 1. Norwich, Climate Research Unit.

Maxwell, B, Mayer, N and Street, R (1997). *The Canada Country Study: Climate Impacts and Adaptation. National Summary for Policy Makers.* Ottawa, Environment Canada.

NOAA (1995). *Natural Disaster Survey Report: July 1995 Heat Wave.* Washington DC, US Department of Commerce.

Rabinovitch, J (1992). 'Curitiba: towards sustainable urban development', *Environment and Urbanization*, 4, 2.

World Bank (2000). *Entering the 21st Century: World Development Report 1999/2000.* New York, Oxford University Press.

7

The water we use and abuse

The demands of increased populations and the desire for higher standards of living have brought with them much greater requirements for fresh water. During the last 50 years water use worldwide has grown fourfold. . . . Increasingly, water stored over hundreds or thousands of years in underground aquifers is being tapped for current use. . . . The availability of fresh water will be substantially changed in a world affected by global warming. (Houghton 1994: 97–8)

[Water] *systems have been sized and structured on the basis of the historical weather patterns; substantial changes in these weather patterns may force cities and counties to make major investments to change them. (ICLEI 2000: 10)*

The year-to-year variability in precipitation [is expected to increase] *almost everywhere* [in the United Kingdom] *even in seasons and regions when mean precipitation amounts* [are expected to] *fall. (Hulme and Jenkins 1998: v)*

7.1 New challenges for water management

7.1.1 Greater uncertainty

The triple aspect of the urban water management problem does not change: sometimes there is too much water, resulting in flood; sometimes there is too little; always there is the need to keep contamination in check. Although we have learned a great deal about urban water management – and billions of pounds have been invested in water-related infrastructure – the future will be challenging, even for cities that have not previously regarded their water supply as particularly vulnerable. Paradoxically, our best hope lies in the fact that we have so profoundly mismanaged our water resources that there is a great deal of scope for improvement. The planet's freshwater supply is already under serious pressure even without the threat of climate change.

The intensification of the hydrological cycle due to climate change will pose serious challenges for the management of water supplies. Climate change injects a huge amount of uncertainty into the water manager's task. Nor will we simply have to adapt to a new hydrological regime, because climate will continue to change as long as we continue to emit greenhouse gases in excess of the ocean's absorptive capacity, and for several decades after that. The increased irregularity of the rainfall regime means that cities will be more prone to droughts as well as floods. More intense rainfall events also have a greater propensity to mobilise contaminants and pathogens on and near the surface; thus the contaminant impact of the 'first flush' of a rainstorm in a city will increase.

7.1.2 Agricultural demand for water

Warmer temperatures also mean that the water demand from agriculture will increase to counterbalance the moisture loss through increased evapotranspiration. Some agricultural scenarios predict yield increases thanks to a longer growing season and carbon dioxide enrichment, *assuming* that additional water will be available. The urban focus of this book means that agriculture is rarely mentioned, yet, in the context of the competition for scarce water resources, agriculture cannot be ignored. This is simply because agricultural water use accounts for two-thirds of total human water use and in many parts of the world, such as southern California, cities and farmers have been in intense competition for water for decades. Furthermore, some urban administrative areas, like Beijing and Shanghai, include considerable amounts of farmland within their boundaries.

The interplay between agricultural activities and urban/industrial activities is also a key factor in controlling the release of harmful residuals. Dangerous pesticides and pathogens can be leaked from farms into the urban water supply, while crops and farm animals may take up urban/industrial contaminants.

7.1.3 New approaches to water management

The water management challenge is increasing because the global population is still growing, it is still urbanising and people are becoming wealthier. Water use goes up as people urbanise and grow richer. We are thus increasing the stress on a resource that we have been mismanaging in the Western world for at least 200 years. Throughout much of that history we have reacted to water shortages by increasing the supply of raw water through pumping groundwater from greater and greater depths, by building more reservoirs and by constructing viaducts and pipes to make inter-basin transfers. Latterly there has been more emphasis on reducing in-pipe losses, recycling water and reducing demand through more realistic pricing. However, this recent tendency is pulling against a much longer tradition of extrapolating an increase in demand from past growth and then making more raw water available for human use.

We can see an example of the tension between these two tendencies in China today. The Beijing–Tienstin–Tangshan region is home to 26 million people in a triangle with 100 km sides, with annual rainfall ranging from 700 mm to 400 mm per year. Growing water demand in this region lies behind plans to pump water 1200 km from the Yangtze Basin, crossing 219 rivers in order to do so. Meanwhile, Beijing is also in the process of building 12 sewage treatment plants and metering houses for water use for the first time.

The basic problem is that we are fast approaching the time when there is no more 'surplus water' to be abstracted from distant river basins. It is no coincidence that this exhaustion of new water supplies is occurring at the very same time that our greenhouse gas emissions are warming the climate and thus bringing in a period of great uncertainty for water management. It is the rapid growth of the human population, based precisely on our ability to increase our throughput of natural resources of all kinds – especially energy and water – that has suddenly brought us to this point.

7.2 Water shortages

At this time the biggest fear is that we will experience water shortages as we reach the limits of the available supply. This is visible evidence of usage exceeding sustainable supply in cities that have so over-pumped the groundwater that widespread subsidence has occurred, as in Venice, Mexico City and Bangkok. In these cases the reasons for the problem are obvious. On the mainland opposite Venice, major water abstractions have been made by heavy industry. Mexico City has a population of over 20 million people in a dry region, and again industrial usage of groundwater has been permitted. In Bangkok the main demand for groundwater has been from agriculture.

In cases like these it is clear that water abstraction on this scale is not sustainable. A basic operating rule should be that no abstraction of groundwater is permitted if it exceeds the recharge rate. Of course, the recharge rate varies considerably from year to year; but the permitted level should be set at a level that will only exceptionally exceed recharge. This target would seem to be *the most elementary precaution* because it is necessary to maintain the most important point of balance between human beings and the biosphere. And yet, even in a complex economy like the United States, this elementary imperative is totally ignored and groundwater is heavily overdrawn.

These examples might suggest that only major cities, in tandem with irrigated agriculture, could produce this kind of crisis. The fact is that – as far as water supply is concerned – we live much closer to the margin than most people realise. Britain, for example, can experience an occasional dry summer that reduces crop yields and stresses livestock, but recently the term 'drought' has come into more frequent use. In 1995–6 reservoirs ran dry in west Yorkshire, normally a well-watered part of the country. Usage was restricted, standpipes were erected in the streets and a fleet of 700 tanker trucks was used for five months to refill the reservoirs (Bakker 2000).

Despite the elaborate system of government surveillance that has been established in Britain to regulate the newly privatised supply of water, important ambiguities regarding responsibility remain. High salaries are paid to senior management, dividends are paid to shareholders, but water may not be delivered to customers. In some cases of failure the water companies have been fined, although the penalties are trivial compared with the profits they earn (see, for example, Ravetz 2000: 134). Given the natural variability of the water supply, it is not always easy to establish guilt if water is not available. Under climate change the determination of responsibility will become even more difficult, for who can be held responsible when the future is so uncertain?

The fact is that, although the UK stays green throughout the year, river basins are very small, by continental standards. This means that – at any point in time – the amount of water stored in the hydrological system is small in relation to annual usage. Small oscillations in rainfall can put the supply in jeopardy at this level of use. The potential for periodic failure is likely to increase with climate change (see section 7.6). It might also come as a surprise to know that in Britain in-pipe leakage rates can be as high as 30%. This, combined with the raw water supply issue, has led the UK Drinking Water Inspectorate to require the supply companies to reduce these preventable losses.

The British Isles has one of the world's most favourable rainfall regimes for the support of a dense urban population and intensive agriculture. Yet even here the signs of stress are apparent, long before the full impacts of climate change have been experienced. Almost every other densely populated country lives much closer to the hydrological edge. Many megacities of the South are rapidly assuming Western levels of consumption of all resources, especially water. All of these cities rely upon a strongly seasonal rainfall regime and hence are much more vulnerable to drought than the cities of north-western Europe which enjoy year-round rainfall.

7.3 River basin floods

It may seem paradoxical that both flood and drought can afflict the same region. But that is frequently the case, as when the 1993 drought followed the 1993 floods in the Ohio–Mississippi valleys. Apart from the intensification of the hydrological cycle expected from climate change, humankind has set in train other processes that are likely to increase the risks of both flood and drought. Deforestation is linked to diminished rainfall, and when heavy rains do occur the lack of forest cover accelerates runoff, thereby increasing the likelihood of a flood. During the May–September 1998 floods on the Yangtze River deforestation was widely blamed for the extent of the damage which resulted in more than 10 million people being evacuated, while the city of Wuhan (population 7 million) was itself in danger of evacuation. The floods on the Rhine (February 1995) and on the Oder (July 1997) incurred billion dollar insurance losses. Over the years these rivers have been straightened to facilitate river traffic; however, this also accelerates the flow of the water and increases the danger of flood.

Although river basin management lies outside the scope of traditional concepts of urban management, cities are squarely at risk from these events. Indeed they have the most to lose. Thus, even though urban governments have no formal control over river basin management, there is an urgent need for many of them to integrate their own hydrological and land use planning with the surrounding regional authorities (Kulkarni 2000). What they also need to do, to lessen the potential impacts, is to ban construction in the flood plain, despite development pressures to the contrary.

The situation is not unlike the policy debate on climate change. Cities have no formal role in the issue because it is in the hands of national governments. Yet they are the most vulnerable parts of the country and also they have the most immediate means at hand both to develop adaptive strategies and to reduce emissions. Similarly, in the hydrological debate, cities need to do more to protect themselves from the risks that are already out there, and they need to do all they can to reduce the level of that risk. However, just as cities have evolved with no thought given to their overall

7.1 Short term protection. The building of levées provides protection in the flood plain from smaller floods but not for major events. This is the town of Hannibal in the 1993 Mississippi–Missouri flood.

Source: Swiss Re. 1994.

7.2 Double vulnerability. The town of Grand Forks, North Dakota suffered both flood and fire in the Upper Midwest floods of 1997. The spring snow-melt overwhelmed the Red River which flooded its banks from Minnesota, through the Dakotas and Manitoba. A fire in the downtown area was cut off by the water, making access for the fire-fighters very difficult.

Source: Partner Re 1997.

energy efficiency, so they have evolved with no thought given to their place in the global hydrological cycle.

7.4 Urban floods

7.4.1 Rising groundwater

Paradoxically, a more insidious flood threat is building up under London because *less* water is being abstracted from the groundwater than was the case throughout the last 200 years. Heavy water-using industries, such as breweries and food processors, have gradually moved out of the city to suburban locations or relocated overseas. Through the nineteenth and twentieth centuries new buildings were erected in relation to the lower water table that then prevailed. Now the water table is rising again and the foundations of many buildings are at risk. The danger is twofold. First, the rising waters threaten the foundations of the buildings and increase their vulnerability to flooding. Second, the ground through which they are rising is usually highly contaminated by decades of industrial and commercial activity. Thus the rising waters are likely to mobilise contaminants and introduce them into the local water supply. The same problem is found in many old industrial cities located in sedimentary basins, such as Manchester and Paris.

7.4.2 Infrastructure floods

A common weakness in the development of cities is that, inevitably, their central areas are served by the oldest infrastructure, which is particularly critical for the management of sewage and stormwater. These older drainage systems eventually become the hub of a much larger drainage area and 'sewershed'. Despite upgrades, central cores of cities are often overwhelmed by storms, especially high intensity convective storms that are likely to occur in the summer, particularly in continental interiors. New developments in the suburbs may contribute further to the flood risk by reducing the amount of land open to absorb rainfall. Construction close to remaining river systems contributes to sedimentation in the channel, thereby reducing flow capacity and hence increasing the impact of floodwaters. These are problems of richer countries and they should be called 'infrastructure floods' because they have been created by deliberate land use and infrastructure choices (Kulkarni 2000).

7.4.3 The insurance (property and casualty) industry and urban floods

Once again, the contradictions arising in the modern urban environment fall on the insurer's desk. In countries like the United States and France the government funds a compensation scheme for victims of floods. In Britain flood damage is usually covered under a standard homeowner's insurance policy.

In Canada compensation is provided jointly by provincial and federal governments, but only for major river basin floods, not those of the urban infrastructure variety. Homeowners' insurance policies will cover accidental pipe bursts and sewer back-up, but *not* damage from floods of the urban infrastructure type mentioned above, which tend to strike the same parts of a town year after year. Insurers maintain that these floods – like pollution – are not accidental. They are predictable and preventable. For many years flooded homeowners languished in legal limbo, unsuccessfully suing urban governments whom they claimed were negligent in stormwater management. After many years of this stalemate the court found in favour of homeowners suing the city of Thunder Bay, Ontario, ruling that the city managers had ignored the warnings of consulting engineers, who had recommended remedial action. The provincial and federal appeal courts upheld this judgment and the city was obliged to compensate the victims. After this ruling, premiums for municipal insurance went up by 50% throughout the province.

7.4.4 Inadequate zoning controls

A number of factors contribute to urban buildings becoming vulnerable to flooding. Changes may be made to the upstream channel, which increase the amount of runoff. For example, sections of the river may be straightened to improve navigation. Such decisions could be taken without any consultation with downstream communities.

Within the city, planners may permit unwise upstream paving over of the land – for a parking lot, perhaps – which may increase the risk of other buildings, downstream of the work, being inundated. Budgetary pressures may reduce the return period of the 'design storm' (see section 4.1) used in the planning of new suburbs, from ten years to five, and even down to two years!

However, the most direct means of increasing urban vulnerability to floods is to permit construction within the known flood plain (Munich Re 1997). For many years such areas may have been left unbuilt, but development pressure over the decades can encourage encroachments, especially where the flood plain is adjacent to high value urban land. There are even cases where flooded buildings have been demolished, only to be replaced by new construction once memories of the flood hazard faded.

7.4.5 Illegal construction

In poorer countries the urban flood risk is exacerbated by illegal housing built on river banks which not only tends to block the channels but also places the inhabitants directly in the path of danger. The December 1999 floods in northern Venezuela had major impacts on Caracas and neighbouring towns where it is estimated that 50000 people were drowned. Some of these problems are very difficult for an urban government to address, such as the overwhelming number of squatters in cities in the developing world, many of whom live illegally in flood-prone areas. However, other problems related to infrastructure management in cities in the richer countries certainly can and should be addressed. They remain problematic because the decisions that contribute to them are taken one at a time without reference to the urban hydrological system as a whole, and without reference to other changes in the river basin that might have impacts on the city.

It should be noted that some cities are susceptible to floods from both sources – the river basin and within the urban area itself. Hanoi, for example, undergoes frequent floods in the monsoon season, and in the past several hundred years it coped with them by detaining the water in numerous small lakes within the city. Unfortunately the recent commercialisation of the economy has encouraged the illegal infilling of the lakes to make space for construction, and in doing so destroys an excellent system for the detention of floodwaters. Furthermore, construction is also taking place on the dykes that channel the river away from the city. The same commercial forces favour construction within the flood plain. These developments could seriously increase the vulnerability of the urban system to flooding, especially under the sea level rise scenario, because the city is less than 3 metres above sea level even though it is 90 kilometres from the coast (Timmerman and White 1997: 221).

7.5 Health and water quality

We cannot live without an adequate water supply – yet water itself is also the most potent disease vector for humans because we ingest it daily and many of the dangers that it might bring are invisible. Cholera, malaria, typhoid and yellow fever – the great killers of the past – have been eradicated from Western cities. However, malaria (for which the mosquito vector breeds in stagnant water) is still an important cause of death in developing world cities, where cholera may also be found. Potent causes of morbidity are the numerous bacteria that cause intestinal infections (such as *E. coli*) which, through enteritis and dehydration, can be fatal for children and other susceptible members of the population. For example, an outbreak of *E. coli* 0157:H7 contamination in Walkerton, Ontario, in May 2000 affected 2000 people in a town with a population of 5000. There were seven fatalities. The enquiry continues at the time of going to press (see http://www. walkertoninquiry.com/).

In addition to water-borne disease we face the threat of ingesting the contaminants that we ourselves have introduced into the hydrological cycle, as described in Chapter 4 – heavy metals, pesticides, solvents, PAHs and so on. The major types of pollutants, other than acids and agricultural biocides, are summarised in Table 7.1. These contaminants pose long term risks to health, although often with a long latency period. More rapid impacts are delivered by water-borne pathogens such as those identified in Table 7.2. Some of these pathogens are found only in tropical and semi-tropical regions; others are found throughout the world. Inevitably, warm weather pathogens will become more widespread with climate change. The apparent resurgence of one of these water-borne pathogens will be described to illustrate the impact they can have on established water treatment systems in a modern city.

7.5.1 *Cryptosporidium*

The elaborate precautions taken to safeguard water in rich countries would suggest that the supply could be guaranteed as safe. However, in recent years, the water supplies of many modern cities around the world have been contaminated by *Cryptosporidium*, a protozoa, which induces gastro-

Table 7.1 Major types of water pollutants

Pollutant	Major sources
Oxygen-demanding wastes	sewage effluent; agricultural runoff including animal wastes; some industrial effluents from food processing, pulp and paper etc
Plant nutrients	sewage effluent, including phosphates from detergents; agricultural runoff, especially nitrates from fertilisers
Toxic metals	mining and associated industries; lead from motor vehicle exhaust; incinerators; urban runoff
Oil	natural seepage; waste from transportation operations; casual leakage from installations
PCBs	sewage effluent; waste incineration and disposal to landfill; electrical installations including transformer stations in cities

Source: Adapted from Jackson and Jackson 1996: 282.

Table 7.2 Examples of water-borne pathogens transmitted to humans through polluted water

Pathogen	Disease
Bacteria	
Salmonella typhi	Typhoid fever
Shigella dysenteriae	Dysentery
Vibrio cholera	Cholera
Escherichia coli	Enteritis
Viruses	
Hepatitis virus A	Infectious hepatitis
Poliomyelitis virus	Polio
Parasitic protozoa	
Entamoeba histolytica	Amoebic dysentery
Giardia lamblia	Giardia
Cryptosporidium	Cryptosporidiosis
Parasitic worms	
Schistosoma spp.	Schistosomiasis (bilharzia)
Ankylostoma	Hookworm

Source: Adapted from Jackson and Jackson 1996: 285.

enteritis over a period of two weeks, and which can be fatal for the young, the old and the immunocompromised. An outbreak in Wisconsin in 1993 left 300 dead and 400000 sick. Mortalities have also been reported from California, Argentina, Spain and South Africa. In 1999 a major outbreak struck Sydney, Australia. There is no treatment for the illness and standard chlorine-based methods of disinfection fail to kill the pathogen.

In the UK outbreaks in Swindon and Oxfordshire in 1989 led to the formation of an Expert Group to report to the national government. Despite the additional precautions that were taken, an outbreak in February and March of 1997 resulted in 345 confirmed cases of cryptosporidiosis in north London, which were attributed to inadequate standards at a treatment station (Gray 1999). What was unusual about the London case was that it 'was associated with drinking water derived from underground strata. Such sources had generally been considered at low risk of contamination by *Cryptosporidium*' (DETR 1998: 1).

This was not such a rare event. Indeed, it was the twenty-fifth outbreak in the UK since 1988. A new standard has now been established and water companies in Britain will be required to sample continuously and test daily for *Cryptosporidium*. There is no agreement on why these outbreaks are occurring now, as the organism evolved long before humans who presumably have been exposed to it over millennia. However, it has been noted that outbreaks often occur after unusually heavy rainfall events, which might carry the pathogen from cattle faeces directly into the groundwater. Heavy rainfall, and the accompanying increase in turbidity, might also compromise the effectiveness of water treatment stations. The intensification of the hydrological cycle due to climate change would suggest that the risk of these outbreaks would increase.

A further contributory factor, especially in North America, could be a tendency to locate intensive livestock feedlots close to urban areas. This possibility suggests that urban water supply needs to be evaluated in the context of changing land use practices in the surrounding rural areas. For example, the most serious health risk following the landfall of hurricane Floyd in North Carolina (September 1999) was the destruction of holding areas for pig wastes. In New York City testing for *Cryptosporidium* and *Giardia* has been in place since 1992 and the results are posted on a website, in response to widespread public concern (see http://www.ci.nyc.ny.us/html/dep/html/pathogen.html). However, despite this precautionary approach it is admitted that tests are inadequate in that they cannot determine if the oocysts are alive or not, and hence whether or not they pose a threat to health. Furthermore they cannot distinguish between various species of the pathogen, only some of which are a problem for humans.

7.6 The impact of climate change

Climate change will have multiple impacts on water supply, water use and water treatment. Although it is possible that some beneficial changes might occur, the most predictable forecast is that water supply will become much more uncertain in almost all regions of the world. The historical record will become unreliable for planning purposes; urban managers and citizens will need to adopt a precautionary approach. The predictions suggest that supply will become more irregular, use will go up, and treatment will become more difficult. Higher temperatures will encourage the development of pathogens, oxygen-demanding plants, and algae in the water, all contributing to threats to water quality. In general:

> the problem of organic pollution in rivers is usually worse in summer months. The rate of water flow is slower and the volume of water carried is less, leading to an increased concentration of organic pollutants. Higher water temperatures decrease the solubility of oxygen and favour bacterial growth, thus exacerbating the problem of oxygen depletion. (Jackson and Jackson 1996: 283)

Irregularity in supply will occur on at least two time scales. Seasonality may become more extreme, such as drier summers and wetter winters being anticipated for southern England (Wade *et al.* 1999: 73). Convective storms (thunderstorms, hailstorms and tornadoes) are likely to become more intense. For example, a hailstorm in Minneapolis – St Paul in May 1998 was cited as 'the most expensive hail loss in insurance history anywhere in the world' (Munich Re 1999: 4). In May 1999, Oklahoma City was struck by the world's first 'billion dollar tornado' – an F5 (the maximum) on the Fujita scale. Both these tendencies – seasonality and intensity – imply that water suppliers will have to increase storage capacity, while urban planners will have to adopt a more pro-active approach to reducing flood risk.

Water use is likely to go up in response to global warming because of the increased demand for irrigation to combat moisture losses from increased evaporation from the plants and from the soil. While it is certain that irrigation water could be used much more efficiently than is presently the case, overall use is likely to increase. There will also be more evaporation losses from reservoirs.

Water treatment will become widespread and more thorough as the pressure to reuse water increases. Cities that currently treat little of their sewage and wastewater will be obliged to do so by necessity. Countries in which standards are already high will see them rise further in response to new threats, as can be seen in the responses in Britain and the United States to cryptosporidiosis.

In the United States there are important regional differences in the changes expected in the pattern of precipitation:

> Climate models predict that global warming will produce increasing amounts of precipitation for most United States regions – with the exception of the southern U.S. from Arizona and New Mexico across the south to Florida, where precipitation levels are expected to decline. These scenarios mean that many U.S. cities and counties will be forced

to make major investment in new water infrastructure to adapt to the new circumstances. (ICLEI 2000: 9)

The same report went on to note that:

certain portions of municipal infrastructure appear especially vulnerable:

- Changing patterns of precipitation can be expected to overload existing storm waste systems in some areas and lead to inadequate fresh water supplies in others;
- Hurricanes, blizzards, ice storms, tornadoes, hailstorms, floods, and storm surges of increasing frequency and intensity can be expected to increase the amount of damage inflicted on roads, bridges, tunnels, transit systems, airports, utility systems, buildings, parks, and other municipal structure;
- Extreme weather events can be expected to interrupt the operation of such facilities, which in many cases are essential to commerce, health, and safety or to place greater demand on them (e.g. electric power systems during heat waves);
- Rising ocean levels can be expected to render substantial existing investments in waste water, levee, and dike systems obsolete. (ICLEI 2000: 9)

Finally the report warned that urban governments in the United States were unprepared for the challenges:

Few if any cities, however, have factored in anticipated stresses on their systems from climate change [author's italics]. Particular threats from changing precipitation patterns include:

- Decreases may cause rivers, lakes, aquifers, and snow packs that supply cities or counties to decline in volume, substantially reducing the amount of water and (where hydroelectric power is important) reducing the amount of electricity available to the area;
- Increases in evaporation from lakes, rivers, and reservoirs that supply a city may also substantially reduce the amount of water available;
- Increases in annual precipitation and extreme weather events – such as intense rainstorms and heavy snow packs – may overload reservoirs, flood . . . control systems, and sewage treatment facilities.

Decreases in annual precipitation would hit the [American] southwest particularly hard, since it is hard pressed to meet current demand of existing populations under current climatic conditions. (ICLEI 2000: 10)

What will make these conditions particularly difficult to handle is that they will not be stable. As global warming intensifies, climate and hydrology will continue to evolve. For example, cities in Canada and the northern United States may have to deal with an unpredictable snow-melt, with

sudden thaws, for a number of years. After that there may be no dangerous winter snow build-up ever again. Similarly, for the big cities in the
north-east, the major storm hazard may shift from winter blizzards to
summer hurricanes and convective storms.

7.7 Conclusion

The process of urbanisation has gradually removed us from everyday
awareness of the actual scarcity of freshwater. Modern technology has suc-
ceeded in importing ever greater quantities of water to our cities to meet
our insatiable demands. The profligacy of our cities has been exceeded only
by the profligacy of our agriculture. Water has been supplied at very low
cost to the consumer, whether a household or a business, to promote health
for the individual and employment for the economy. Our recourse to
chemical-dependent agriculture and our lax supervision of the disposal of
industrial contaminants has degraded the water that is available, making it
expensive or infeasible to purify and reuse. The richer the society the more
water we use and the more complex the mix of contaminants that we flush
into the water supply.

Despite elaborate surveillance systems our urban water supplies are still
vulnerable, even to well-known parasites, such as *Cryptosporidium* which
has recently infected supplies in America, Europe and Australia. It is sus-
pected that the resurgence of this particular problem is related to intense
rainfall events that flush it rapidly from farm animal wastes into water
treatment plants.

Such rainfall events are just one, minor example of the changes to the
hydrological cycle that we expect to come with climate change. The
expected intensification of the cycle will challenge water managers with
the twin threats of flood and drought. In the short term we can respond to
this challenge with increased storage capacity, both in reservoirs and in the
ground. Groundwater storage will become even more important as reservoirs experience more evaporation losses. Increased detention on the urban
surface and increased groundwater storage will mitigate the urban flood
hazard.

All of these are simply stopgap measures as we respond to decades of
neglect and the new challenge of climate change. Survival through the
medium term requires more fundamental change in the human role in the
hydrological cycle. First we have to recognise that our cities are part of that
cycle. Urban floods still take us by surprise even though we have expressly
created the conditions to make floods more likely, by neglecting our infrastructure and by paving over most of the ground in our cities. Second we
have to become aware that we are close to the limit of what we can sustainably abstract from the freshwater portion of the global cycle. Water is
not another throwaway consumer product – it is the thread of life for which
we have no substitute.

As we instigated the agricultural and industrial revolutions we sluiced our crops with water, we ran huge quantities of water through our factories and used it to cool our power stations. Some householders poured it over their gardens and their automobiles. Others filled up their swimming pools and watered their golf courses. For this we paid very little. In the future we will have to pay considerably more to rehabilitate and conserve supplies. Then we will learn to use less and to use it more carefully.

7.8 Websites

1. City of New York. Department of Environmental Protection. Water Supply Report 1999: *http://ww.ci.nyc.ny.us/html/dep/html/wsstate.html*

2. City of Toronto. Annual and quarterly water quality reports: *http://www.city.toronto.on.ca/water/*

3. Environment Canada. Water Efficiency/Conservation: *http://www.ec.gc.ca/water/en/manage/effic/e_weff.htm*

4. Foundation for Water Research (UK): *http://www.fwr.org/index.html*

5. ICLEI (International Council for Local Environmental Initiatives).

   The Water Campaign: *http://www.iclei.org/water/*

6. Sierra Legal Defence Fund: 'Waterproof – Canada's Drinking Water Report Card': *http://www.sierralegal.org/clear/SierraRprt7.pdf*

7. UK Department of the Environment, Transport and the Regions. Drinking Water Inspectorate: *http://www.dwi.detr.gov.uk/*

7.9 Further reading

Houghton, J (1994). *Global Warming: The Complete Briefing*. Oxford, Lion Publishing.
Jackson, A R W and Jackson, J M (1996). *Environmental Science: The Natural Environment and Human Impact*. London, Longman.
Ravetz, J (2000). *City – Region 2020: Integrated Planning for a Sustainable Environment*. London, Earthscan.
Timmerman, P and White, R R (1997). 'Megahydropolis: coastal cities in the context of global environmental change', *Global Environmental Change*, 7, 3: 205–34.
Wade, S, Hossell, J, Hough, M and Fenn, C, eds (1999). *The Impacts of Climate Change in the South East: Technical Report*. Epsom, W S Atkins Ltd.

Part IV

Health: restoring urban ecosystem health

8

Restoring urban land to productive use

From a regional land use perspective, efforts to preserve rural communities and productive farmlands, wildlife habitats and biodiversity have reinforced the need to limit urban sprawl by encouraging brownfields and infill development. . . . Chicago and many other municipalities have demonstrated that environmental contamination is not an insurmountable liability but rather may be used as a tool to spur economic development in former industrial areas. (Trumbull 1999: 295)

More thought is now being given to reviving old urban villages and to constructing new ones for the 21st century. The focus is on creating spaces with many functions, mixed neighbourhoods where people can live, work and play, where greater density creates a pattern of rich urban experience. (Girardet 1999: 49)

8.1 Reducing our ecological footprint

Until now people in the richer countries have expanded their use of resources without any regard for the environmental constraints of which we are belatedly growing aware. In poor countries, widespread poverty is the immediate constraint to more intense use of resources. If they become rich their people will – following current trends – embrace the Western lifestyle with gusto. Religions and cultures that discourage personal accumulation of wealth (whether communist, Buddhist or animistic) are broken down by the powerful images of Western materialism, reinforced by radio, cinema, television and now the web. In the final five years of the last century China moved swiftly from Maoist control to vibrant commercialism. The baggy blue suits are gone; mobile phones and motorbikes are in. The same pattern can be seen throughout the South, *especially in the cities*.

The rate of change of those poorer countries that are escaping from material poverty is a warning to those of us who still cling to the assumption that environmental constraints can always be matched by more efficient technology. It is imperative that we understand our recent history and reduce our ecological footprint, if only to prove that it can be done, before it is too late. The key to a future with a more appropriate footprint lies in our cities.

Although we cannot see our individual footprints on the landscape we can see their urban manifestation as it spreads out over the countryside, clogging its arteries with traffic and leaving derelict land at its heart. We must restore our cities to productive use, by which I mean *biologically* productive use, as well as economically productive use. At the same time we must restore density to the centre so that the system can provide the *accessibility* that was its original *raison d'être*. How can this be done? Are green space and residential density not competing concepts?

The solution to this apparent paradox is the astonishing level at which we have wasted resources – including land – until now. Indeed, the proliferation of derelict land at the centre of some Western cities (plus the vast amount of land committed to the automobile) means that residential densities are quite low in the centre too. The density that we need lies somewhere on the continuum of the density found in the late nineteenth century slums of London and New York and the ultra-low density of the automobile-dependent suburb. If we maintained an average density of 5000 people per square kilometre through the urban system, then the dynamics of urban travel could be revolutionised in favour of cost-effective, attractive public transport.

We need a mix of density and green space throughout the urban system. This does not translate into a park surrounded by apartments. Nor does it mean layouts mimicking the new urbanism of glossy architectural magazines. We need density to keep the heart of the city alive and to make district heating and cooling economically viable. We need density in the suburbs to make public transport viable. Right now – at least in North America – people living in the suburbs may find that they have little choice but to own one automobile for every adult working outside the household. In addition to maintaining urban density, we need green space for recreation, flood control, food production, air quality and aesthetics. Greater diversity and less throughput are the twin goals.

8.2 Reducing throughput

8.2.1 Individuals, households and other actors

'Throughput' was the term devised by Daly and Cobb (1989) to describe the quantity of materials that we use to maintain the human economy.

The concept is similar to Wolman's notion of 'metabolic intensity' and Wackernagel and Rees's model of our ecological footprint. It tells us how much of each kind of resource we use to carry out our various activities such as eating, washing, working and playing. As individuals, our choices may be constrained by circumstances, such as a child being brought up in an automobile-dependent, low density suburb, or a person looking for work in a city with an inadequate public transport system. Such circumstances can be modified only by a collective decision. Shoppers have choices as to how much packaging they wish to buy with their purchases, but that choice will be constrained by availability. Householders may wish to recycle, but the degree to which they can do so partly depends on the collection of recyclable materials by the local authority. It is easier to compost if you have a garden, however small. In the UK, for example, there appears to be little pressure to reduce unnecessary packaging either, so households produce far more waste than they need to.

Little is known about the dynamics that motivate individuals in their various roles – consumer, householder, citizen – to take action to persuade other actors in the system to respond to their preferences. Likewise, little is known about the interplay between the various actors, such as those same individuals, manufacturers, retailers, city politicians and city planners.

When all of their actions are aggregated the difference in performance is considerable. It has been noted that London recycles only 4% of its solid waste, whereas Vienna manages 40%. There is no doubt that there is a huge potential for reducing the solid waste output from a city that has not already embarked on this path. An 80% reduction is easily within the reach of a household with a garden or other access to composting, plus the kerbside collection of paper, glass, plastic and metals by the local authority. Some authorities also collect garden waste from those households that cannot absorb all their own.

The phrase 'absorb their own' touches on a key concept that underlies any urban waste reduction initiative. Some cities commendably set themselves the target of absorbing 'their own' wastes without knowing how they would be able to achieve this either in terms of disposal sites or agreeing on who 'they' are. Do 'they' include commuters who work in the city and live elsewhere? What about tourists, students and other temporary residents? The drawing of the boundaries for waste management is important. However, a mixed land use would provide more opportunities. For example, excess garden waste and other compostable materials can be mulched for use in city parks.

Just as corporations have set themselves a target of 'zero waste' so could cities. It might take a while to dispose of particular items, but once the difficult items have been identified the responsibility could be passed back to the producer, as has already happened in Germany. It is true that some recyclables cost more to collect than they fetch at sale, the prices of recyclables can be very volatile and markets are still fragile. But in many cases this is simply because too many cities are working in isolation rather than forming regional alliances to develop markets and to reduce waste. Most of what we currently send to landfill is being sent away because the problem has never been analysed for what it is – an avoidable waste of time, energy and materials. Following the imposition of the Landfill Levy, some cities in England now pay so much for the removal of solid waste that they have shut down libraries and other public facilities to meet the expense! Why not take concerted action to reduce the amount of waste instead?

8.2.2 The construction industry

The construction industry is reorganising itself to reduce waste by reusing old materials on site. In Britain, reuse of building materials has been encouraged by the Landfill Levy. This is especially important for rubble (bricks, cinder blocks, concrete), which can be sorted on site and reused as foundation material for the replacement building. In the United States it has been estimated that 70% (by volume) of residential demolition waste could be diverted from landfill (Table 8.1). Demolition waste accounts for nearly half of all waste from construction and demolition (Table 8.2).

Once the options for the reuse and recycling of these building wastes are better understood, a similar lesson can be drawn to the one learned from remediating contaminated land: *the costs of prevention are much lower than the costs of clean-up*. Thus the more sustainable future for the construction industry lies less in recycling old materials than in designing new buildings for which life cycle analysis is an essential feature (Graedel 1999).

8.2.3 Sustainable housing

New types of 'sustainable housing' are being developed (Dougan 1998). In this context, 'sustainable' does not imply that the house will last for ever, only that the materials, energy and water use will contribute to the sustainability of the human occupation of the planet by reducing the throughput associated with the construction and use of the building. Again, if we set ourselves different goals some quite dramatic changes may be achieved. Even in (what used to be) damp and cloudy Britain many houses have enough solar photovoltaic energy potential to provide all their own power needs and sell excess power to the grid. In Toronto – which is colder, but sunnier, than southern England in the winter – there already exists an 'unplugged house', which supplies its own power and water needs, the water being stored from rainfall, used, and then purified and reused (Canzi 1998). Furthermore, it is imperative that this solar energy potential be harnessed for air conditioning. Otherwise rising temperatures will inexorably lead to rising power demand for cooling, which – under current technology – will be met by burning more fossil fuels. This feedback effect would create a very dangerous spiral of ever-increasing energy demand. The July 2000 heat

Table 8.1 Residential demolition waste assessment in the United States (typical site)

	Volume (cubic yards)	Weight (tonnes)
Materials diverted		
Reuse and resale:		
Framing lumber/sheathing	49	8
Brick	12	17.9
Hardwood flooring	7	1.1
Stair units/treads	4	0.4
Windows	2	0.3
Reuse and donation:		
Tubs/toilets/sinks	3	0.7
Doors	3	0.4
Shelves	0.5	0.1
Kitchen cabinets	1	0.2
Recycle:		
Rubble	88	61.6
Metals	13	2.3
Asphalt shingles	10	3.5
Diversion subtotal	192.5	96.5
Materials landfilled		
Plaster	48	21.6
Painted wood	16	4.2
Rubble	7	4.9
Landfill subtotal	71	30.7
Diversion rate	73%	76%

Source: Based on Yost 1999: 181.

Note: Diversion rate based on common building material densities.

Table 8.2 Estimated building-related construction and demolition debris generation, in the United States, 1996, by percentage

	Residential	Non-residential	Total
Source			
Construction	11	6	8
Renovation	55	36	44
Demolition	34	58	48
Totals	100	100	100
Percentage	43	56	100

Source: Based on Yost 1999: 181.

wave in the USA's south-west demonstrates the linkage between ambient temperature and energy use:

> With air conditioners running full blast, electricity demand soared to an all-time high. Customers of Reliant Energy Houston Lighting and

Power used 14.7 million kilowatts, breaking the utility's record of 14.6 million kilowatts, set last August 20th. . . . TXU Electric and Gas said Dallas electricity demand was expected to top the daily record of 21.7 million kilowatts, set last August 26th. The heat also forced dozens of cities, including Austin and San Antonio, to impose restrictions on water use because of falling lake and river levels. (http://www.msnbc.com/news)

Right now almost all our urban roofs are 'dead space', because they neither grow plants nor harness solar power for heating and electricity. On the contrary, in the winter, poorly insulated roofs lose heat that then has to be replaced by burning fossil fuels. In the summer the roofs absorb heat which – throughout North America – we then combat with air conditioning, while the air conditioners generate their own waste heat which we vent to the street. From a thermal perspective, rich human beings are not very smart.

8.3 Density, proximity and variety

One of the fundamental errors of twentieth century urban planning was to establish great swathes of townscape with a single land use – industrial, commercial, residential. It was done for good reason in order to separate housing from polluting industry. However, this reliance on mono-functional zoning – a 'spatial solution' – has proved much costlier in the long run than resolving the problem at source by obliging the producers to cut pollution. Spatial separation prepared the way for the endless spread of low density suburbs, without much retail support or ready access to work and services. In a sense, this development threw away the whole notion of proximity, or accessibility, which is the *raison d'être* for urban centres.

As Herbert Girardet observed in the opening quotation to this chapter, we must now undo this error by making urban land use more diverse again. We must re-create urban neighbourhoods in which people can live, work, shop and socialise without having to get into a car if they want to buy a newspaper or a bottle of milk. It does not mean that everyone will be able to walk or cycle to work, but at least they will be able to live more of their life in one locality and not spend a significant proportion of their day travelling to somewhere else.

In many Western cities people are returning to live in the city and the authorities are facilitating this move by re-zoning industrial and commercial land for residential use. Houses are being put into small infill sites that help to increase residential densities. In Britain waterfront sites are especially valued for redevelopment and many derelict industrial canals are enjoying a new lease of life in urban centres.

Some of these trends may be dismissed as gentrification, and some of the returning rich may indeed be dispossessing lower income renters who

8.1 Increasing residential density. This is a modern infill very close to the centre of Oxford. These flats are located attractively by the river, albeit on a former commercial site that might once have been consigned to institutional use or parking.

Source: Sue White.

8.2 Preserving residential density. In the 1960s downtown Toronto was about to be 'modernised' with freeways and high rises. A fortunate combination of circumstances held the modernising forces at bay while residents renovated these late Victorian streets.

Source: Sue White.

are then displaced to more remote, cheaper parts of the urban system. However, such dispossessions are only part of a much broader 'return to the city' which should eventually provide system-wide benefits for all income groups. In cities where the authorities are aware of the broader trends, efforts are made to provide housing that mixes income groups together. This is the antithesis of the zoning regulations of the post-war period when income groups were very finely segregated by housing lot size and price.

There is some resistance to the movement towards 'intensification' to increase residential densities, because some people fear that it will change the 'character' of a neighbourhood by making it more crowded. However, changes can be made to housing units that make very little visible impact (Sewell 1993). While higher densities can make public transport viable and may enhance a sense of neighbourhood by making human beings more visible on the street and thereby contributing to personal safety.

There are encouraging signs that people are beginning to wake up to their ecological potential once they realise that they are still part of the physical world. There is a growing network of 'eco-communities' that is trying to reattach itself to the environment. In the 1970s such people were a tiny minority who usually established remote rural settlements. Now these communities are much more diverse and are sometimes found within the urban system (Barton 2000). This is not a movement that seeks escape, but one that is committed to rebuild. In North America we are seeing the return of farmers' markets to the city streets. In Europe people were lucky not to lose this phenomenon. The fact that these small enterprises can recolonise the city and sell directly to the public is a very hopeful sign. The fact that the authorities have allowed this development is even more remarkable.

Even some greenfield developments try to emulate the densities of the pre-automobile age, the most famous in England being Poundbury, which is an extension to the market town of Dorchester. Houses are modest in size and each is built as an individual unit, while fitting one of the approved vernacular styles and using local materials. The intention is that residents will be able to walk to shops and services, and some to their places of employment. The houses do have garages but they are tucked away in lanes to the rear of the house, and do not intrude on the streetscape. The site is not yet built out, so it is difficult to judge the final outcome. Critics may want to write Poundbury off as another 'new urbanism' pastiche for the wealthy, but it may well prove to be the first visible example of a return to environmental harmony in an urban setting. Despite the critics, sales are brisk.

8.4 Improving the modal split

Like our use of water our use of urban land is so inefficient that many opportunities for improvement are available. Our legacy of mindless waste dates from the most rapid phase of suburbanisation of Western cities which happened in the 1950s and 1960s. At that time petrol was extremely cheap and energy conservation was not considered in the design of cities, nor of the automobile itself. Designers who belonged to the modernist movement assumed that the emerging, motorised way of life in the United States would become the norm for all those countries that could afford it. In fact, indicators of 'progress' were taken purely from the quantity of goods that

8.3 The new tradition. Poundbury in Dorset is still under construction. All buildings on the site must follow traditional housing design and use local materials. Cars
are permitted, but they do not dominate the streetscape.

Source: Sue White.

people owned – especially cars, televisions and so on. The combination
of cheap suburban land and cheap petrol produced a housing 'solution'
that required an automobile for every kind of trip – work, shopping and
recreation. It became the norm to spend an hour or more each day driving
to and from work. Many of the urban environmental problems we are
struggling with today are the direct outcome of this outdated and wasteful
paradigm for urban life.

   As outlined in the previous section, higher residential density provides
the key to improving the modal split, which means getting people out of
their cars and back on their feet, bicycles and public transport. Because land
use patterns scatter people's destinations all over the map they are drawn
to the most flexible mode of personal transportation, which for a growing
majority is the automobile. However, as long as most workers are trying to

8.4 A pedestrian town. It is possible to walk around Poundbury. Once the site is built out it is hoped that shops, services and employment will furnish a fully pedestrian environment.

Source: Rodney White.

get to approximately the same place (the city centre) at the same time (9 am) automobile dependence will remain highly inefficient, in terms of energy wasted, emissions produced and time spent. The more people who live near their work, and the more people who use public transport, the less intense the crush will be.

Making automobile users pay the full cost of their mode of transport can also encourage improvements to the modal split. Table 3.4 suggests that in the UK in 1993 motorists paid only about 30% of that full cost once externalities such as accidents and the health impacts of air pollution were included. The price of petrol has increased in the UK since then. However, in the United States there is still fierce resistance to price increases and US$2 per gallon is still seen as unacceptably high, yet this is less than half the current UK price. Significant price increases entail unacceptable political risks for politicians. Furthermore, price increases alone will not produce behavioural changes on the scale required to slow global warming. This can only be expected from changes in land use and major improvements in public transport (in the next chapter see section 9.4, 'The transportation challenge').

8.5 Redevelopment and reuse of brownfields

Unused urban lands are symptomatic of our failure to make the most of our opportunities while living within the environmental limits. These sites enjoy a locational advantage over sites on the periphery, while infrastructure services, such as water and sewerage, transportation, and communications, are already available. Not only are they a wasted asset, their lack of use translates directly into lost employment and lost property taxes. Such negative factors may spread to neighbouring lots. They may be used for the dumping of waste materials and other illegal activities. It has been estimated that there may be as many as 500 000 such sites in the United States alone (US General Accounting Office, quoted in Trumbull 1999: 296).

The reasons for this type of situation were introduced in Chapters 2 and 5 where it was stressed that it is not so much the physical costs associated with clean-up that deter development, but the lack of a framework for co-operation between the public and private sector stakeholders. No developer is going to invest in a brownfield site if that means assuming an open-ended risk. Nor will a developer invest in a brownfield site if it is cheaper to build and operate on a greenfield site. Thus the context of the investment decision must be approached in a comprehensive fashion. In Chicago, the Brownfields Initiative has done exactly that (Trumbull 1999). At the meeting that launched the Initiative it was found that a major barrier to redevelopment was simply the lack of communication between the parties that had to be involved in a successful programme, including the local community.

Simply substituting the term 'brownfield' for 'contaminated land' is a step in the right direction as it poses a challenge, rather than finding a problem. In this spirit William Trumbull offers 'an operating definition for a brownfield' as 'a vacant or under-utilised property passed over for redevelopment due to real or perceived contamination'. To which he adds: 'a brownfield is a real estate and economic development issue, complicated by environmental costs and liabilities as well as acquisition and development problems' (Trumbull 1999: 296).

Part of the problem lies in a failure to appreciate how much the economy has changed since brownfield sites were originally used and contaminated, commonly a hundred years ago. Environmental standards are much higher now, because then there really were no standards at all. Industry is now much less polluting and set to become even more so. The industry benchmark for some operations is now zero emissions, and zero waste. For example, Joseph Romm quotes the chairman of Interface, Inc, a manufacturer of carpets and carpet fibre as saying: 'There will be zero scrap going into landfills and zero emissions going into the biosphere. . . . Our company will grow by cleaning up the world, not by polluting or degrading it' (Romm 1999: 181). On this basis some industrial activity could be re-established on suitable urban sites without returning to the environmental problems of the past. However, given the nature of the post-industrial economy it is more likely that a new commercial use would be in the service sector, such as distribution and communications. Other uses could include residential redevelopment, community services, urban gardens and parks.

The first step in reversing the brownfield problem is to recognise that unused urban land is a wasted asset. Brownfields should be regarded as an anomaly in a healthy urban system, not an inevitable symptom of decline. The second step is to bring the interested parties, including the local community, together to examine the options, including the possible future uses, the appropriate standard and technique for remediation, and the apportionment of cost and any future liability.

None of this means that the days of the brownfields are over, because some are unused for reasons other than the problem of contamination, such as regional economic decline. In cities like Liverpool, which has lost 30% of its population over the last thirty years, demand for most empty sites is weak or non-existent. In cases like this only public intervention over the long term can restore the urban landscape and economy.

8.6 Energy from waste and biomass

During the 1970s most of us who lived in Western cities assumed that technical problems must have technical solutions. Thus we were not surprised to learn that the solution for the urban solid waste problem was to burn it, and thereby provide ourselves with cheap energy for the grid. Municipalities bought incinerators and the problem was about to be solved. Unfortunately the contaminants in the waste stream could become potentially dangerous as particulates leave the incinerator chimney, and also when left as residue in the ash. Despite arguments that performance could be improved to eliminate the risk, incineration went out of favour and in many Western countries incinerators have been closed down.

However, the methane that is produced by decomposing waste in landfills (mentioned in Chapter 5) can be burned to produce electricity and this is now being done in a number of municipalities. In Britain, power producers are buying landfills precisely to obtain rights to the methane so that they can burn it to meet part of their Non-Fossil Fuel Obligation. In Toronto the conversion of landfill methane to electricity 'generates over $2 million in revenue for the City' (Loughborough 1999: 124). This is not a renewable source of energy, but at least it is disposing of a residual problem in the landfill. In the longer run this kind of operation should become obsolete as the amount of material going to the dump is dramatically reduced.

There is a longer term future for energy from biomass, both from production wastes such as pulp and paper residues and pig slurry, as well as from crops grown for that purpose, such as ethanol from sugar cane and woodlots. The heat content of these fuels is less than fossil fuels (Table 2.5) but if planted on a sustainable basis they will be carbon dioxide neutral. Once we develop an accounting methodology that will attribute costs to the impacts of climate change it should soon be clear such fuels are much more valuable than their heating value alone would suggest. If we use biomass for fuel then we are on a much safer ecological path than simply planting trees to sequester carbon to offset the carbon dioxide emissions from the burning of fossil fuels. That kind of approach is a one-time only offset which can be maintained only if the forest is managed sustainably. Additional offsets for additional fossil fuel combustion will require more land for additional forests.

There is no reason why urban land cannot be used to demonstrate this important principle. If each city maintained visible renewable energy projects then its citizens would become more quickly aware of the bio-geochemical cycles of which they are part.

8.7 Naturalising urban systems

Despite the ambitions of planners and architects it is rarely they who design and build cities. Today most of that work is done by engineers, each working in their allotted sub-discipline, solving their own part of the urban problem. Apart from the occasional 'New Town', no town is designed as a whole. Instead a town evolves as one 'solution' is added to another. Widening roads and building elevated freeways solves traffic congestion. Channelling rivers and putting in bigger drains solves flooding. Pavements replace muddy pathways. All of these solutions belong to the landfill class of solution – the problem has been relocated, not solved. We are now re-examining some of these solutions from the past to see if what they replaced might be more effective.

We have to reconsider what we mean by 'efficiency', and add to it the goal of productivity, that is biological productivity. The trick is not to replace Nature with inert matter but to learn to live with Nature (McHarg 1969; Owen 1991). The implications of this viewpoint will be examined in the next two chapters with a focus on water and air. The remainder of this chapter will examine the implications from a land use perspective.

8.7.1 Urban forestry

There are multiple benefits to be gained by developing the urban forest, both stands of trees in parks and individual trees in private gardens and along the road. In summer they provide shade and in the winter they provide shelter from the wind. A well-developed tree canopy can reduce the urban energy demand for space heating and cooling significantly (Akbari and Taha 1991; Akbari *et al.* 1992; Duffy 1999). Trees also enrich the urban niche for birds and other wildlife, some of which may be more welcome than others, such as foxes, racoons and skunks. As noted above, part of the urban forest can be used to provide energy for the city, if only as a demonstration project.

It might be argued that urban land is too valuable for gardens or any type of urban farming, but gardens need not be confined to the street. Vertical gardens can be developed from rooftops and window ledges to pro-vide cooling to the buildings' surface. Flat rooftops are also a vast unex-ploited resource for urban gardening. There are problems of pollution and health standards, problems of the load-bearing capacity of roofs that were not built for this purpose, and so on. But solutions will be sought only if a new objective is defined for the city, that being to make the city a productive part of the biosphere, rather than simply a jumble of conflicting and polluting uses.

8.7.2 Green infrastructure

The notion behind green infrastructure is that much of the 'hardware' of the urban system could be designed to operate in a more natural fashion (Husslage 1997). As mentioned in the previous chapter, it is important that we increase the ability of the urban surface to permit the infiltration of precipitation. The most obvious target for such rehabilitation is the parking lot, the very symbol of urban dysfunctionality. Most parking lots are built simply to maximise the number of parking spaces; drainage is designed to take the rainwater off into the drains as quickly as possible. The extensive, paved surface contributes to the heat island effect. Only in the wealthier suburbs is some effort devoted to providing shade trees. The higher temperatures that will come with global warming should make this shading function more important.

A combination of porous paving, trees, shrubs and climbing plants will:

- improve ambient air quality
- sequester carbon dioxide
- save energy (reduced heating and cooling needs in adjacent buildings)
- reduce the urban heat island effect
- decrease stormwater runoff (Jankovic and Amott 1998).

8.8 Conservation of historic buildings and districts

One of the lasting side-effects of the modernist movement was the widespread destruction of older buildings and neighbourhoods to make way for universalist buildings and highways. In some bizarre way it seemed to be assumed that this convergence on sameness was a sign of progress, as if we shared the same destiny to spend our lives in boxes of steel, concrete and cement. The destruction caused by the Second World War provided an opportunity for the rapid proliferation of very ugly structures, no doubt justified by the argument that they could be assembled quickly and cheaply.

This is not the place to investigate this sad episode in urban history other than to note that an important aspect of the 'variety' referred to in section 8.3 is *variety in time*. A sense of our historical roots can most readily be enjoyed if some part of the past is still visible around us, and where possible, in active use. Much of the work on historic conservation appears to be geared to tourism or sometimes to a more lofty goal of world heritage, but the greatest benefit is for the everyday enjoyment of local people who have the opportunity to understand that they live in time as well as space. The way we do things today – such as sitting in traffic jams and creating mountains of waste material – is not the way we always lived, and thus not the way we have to live for ever more.

Part of the anomie of modern urban life comes from the dreadful sameness of much of the modern city whether built by the forces of capitalism or communism. At times this trend has been forcefully rejected as in the case of the rebuilding of the Old Town of Warsaw after its total destruction in 1945. It could have been argued that a fresh start should be made or that the old buildings were irrelevant to a post-war city, but instead the past was rebuilt and is actively in use today, by local people as well as visitors.

Without a sense of history there is no sense of time in the city other than the diurnal flows of traffic and commerce. Yet a city can display its richness only if its history is visible and the variety of urban life is on display and in active use.

8.9 Conclusion

Cities evolved and continue to grow to meet a variety of human needs that have one feature in common – accessibility. Yet the Western city is in danger of failing in this primary objective because the preferred means of transport – the automobile – cannot operate at the volume that people expect. In response to the consequent congestion highway engineers destroyed the

8.5 A new view of Warsaw's Old Town. It is said that at the end of the Second World War not one brick stood upon another in the rubble of the Old Town. Yet rather than replace the square with another cement tower block the people rebuilt it exactly as it had been. Today it is once again a vibrant centre for visitors and local people alike.

Source: Sue White.

functionality of the city core. As Jane Jacobs observed in her landmark book *The Death and Life of American Cities* (1961), the highways flattened some neighbourhoods and isolated others. Traffic jams are no substitute for urban community life, however many toys you carry in your car. As a by-product of this functional failure we have polluted the air and changed the composition of the atmosphere. It is time to return the notion of 'city' to the drawing board.

For this and many other reasons we must reduce our level of resource use and waste generation. If we are to continue to live in cities *and* reduce our ecological footprint we must first restore functionality to the city by ensuring that people have access to the jobs and services that the city can be expected to provide. Easy, affordable access means an end to mono-functional zoning and an end to the abandonment of contaminated land. We must ensure that cities can function without exporting their problems to the countryside. We must restore cities to a sustainable role in the biosphere.

8.10 Websites

1. City of London. The Liveable City – a handbook to improve sustainability: *http://www.cityoflondon.gov.uk/environment/hbookfrmd.htm*

2. Environment Canada, Adaptation and Impacts Research Group at the University of Toronto. 'Adapting Urban Areas to Atmospheric Change': *http://www.msc-smc.ec.gc.ca/airg/vertical_gardens.htm*

3. New York City – Department of Environmental Protection: *http://www.ci.nyc.ny.us/html/dep/home.html*

8.11 Further reading

Daly, H E and Cobb, J B (1989). *For the Common Good*. London, Green Print.

Girardet, H (1992). *The Gaia Atlas of Cities: New Directions for Sustainable Urban Living*. New York, Anchor Books published by Doubleday.

Girardet, H (1999). *Creating Sustainable Cities*. Totnes, Green Books for the Schumacher Society. Schumacher Briefing No 2.

Jenks, M, Burton, E and Williams, K, eds (1996). *The Compact City: A Sustainable Urban Form*. London, E and F N Spon.

Ravetz, J (2000). *City – Region 2020: Integrated Planning for a Sustainable Environment*. London, Earthscan.

Romm, J J (1999). *Cool Companies: How the Best Companies Boost Profits and Productivity by Cutting Greenhouse Gas Emissions*. London, Earthscan.

Sewell, J (1993). *The Shape of the City: Toronto Struggles with Modern Planning*. Toronto, University of Toronto Press.

Trumbull, W C (1999). 'The Chicago Brownfields Initiative', in *Reshaping the Built Environment: Ecology, Ethics and Economics*. Ed C J Kibert. Washington DC, Island Press: 295–309.

9

Clearing the air

The solutions are not as drastic or as undesirable as some might feel. It is not necessary to mandate a lot of things to achieve significant per capita reductions [of carbon dioxide emissions]. *Much progress could be made if we thought that the issue was so serious that we HAD to reduce emissions. Then it would be possible to push much harder by implementing measures like gasoline taxes. . . . It might be difficult, but if we thought that was our only choice, if our whole system was going to change without taking this action, we would do it without hesitation. (Ogilvie 1999: 131)*

Developing a carbon market for domestic consumers is, at least initially, also about awareness and education. The majority of private individuals in the UK do not realise that their activities result in carbon dioxide releases, or that these are the primary cause of climate change. (Fawcett et al. *2000: 75)*

9.1 An integrated approach

Although the diversity and complexity of the problems we have created by burning fossil fuels are daunting, the task need not be overwhelming. The fact that the problems range from local air quality issues to global climate change should not be an obstacle to an integrated management approach. Because our problems have a common cause they can be dealt with as part of an integrated recovery programme. Our problem – as noted in Chapter 1 – is that modern society is structured to deal with the symptoms of our disorder, not the cause.

As with the solid waste problem we must ask the question: 'What happens to the residuals next?' And we need to keep asking that question until we are satisfied that the residuals are inert and pose no further problem. For some residuals, like carbon dioxide, we will not reach a point where there is no further problem – at least not until the decades have passed while we eliminate further emissions, plus another hundred years for the last part of the atmospheric overload to be absorbed back into the oceans. For those residuals for which the problem time horizon is extremely long, or even infinite, we must go backwards through the R–A–R chain of events to either change the nature of our activities or change the types of resources we use to support them.

9.1.1 Initiating the recovery process

We begin this recovery process – as in the solid waste problem, again – by monitoring what we are actually doing. We need a picture of all our activities and the problematic emissions for which they are responsible. For example, Adur District Council, in Sussex, has begun to do this. Concentrations are recorded monthly for sulphur dioxide, nitrogen oxide and ozone (Table 9.1). Adur, with a population of less than 60 000 people and with an area of about 10 000 hectares, is 'one of the smallest district councils in England' (Adur District Council 1999: 1). Its leaders have to work with the usual plethora of regulations, taxes and budgetary pressures from higher levels of government. Despite this, the council has taken the first coherent steps to restore air quality and improve the environment in the locality.

In physical terms it is not difficult to conceptualise an integrated approach to the problem; it is much more difficult to do so in political terms. Responsibility for atmospheric emissions is fragmented, and in some critical areas like greenhouse gases, not yet in existence. Local authorities have some monitoring and enforcement responsibility for air quality issues, but a higher level of government usually sets the standards. Local authorities have highways running through their area of jurisdiction from which vehicles emit greenhouse gases. We have not begun to decide how to attribute responsibility for emissions – to the jurisdiction where they are emitted, to the locality where the driver lives, where the vehicle owner lives, or where the fuel is purchased? How do we bound the problem? Later on in the process the question of the attribution of responsibility will have to be faced.

Ultimately, emissions that affect air quality can be managed only on the basis of air-sheds – the regional air masses that affect a locality. This is recognised in Ontario where an emissions trading scheme among electricity generators is being established to reduce sulphur dioxide and nitrous oxide. Trading may be permitted for the air-shed, meaning up to 1500 km upwind of Ontario, including American states like Ohio and Tennessee, but excluding New York 'as most pollution from that state is blown over the Atlantic. Also permits from plants further away from Ontario would be

Table 9.1 Air quality monitoring by Adur District
Council, Sussex, UK, 1998–9

Monthly average	SO$_2$ µg/m^3	NOx ppb	O$_3$ ppb
April	10.84	19.70	30
May	16.01	21.50	30
June	12.98	18.50	28
July	8.14	22.90	24
August	9.16	28.10	25
September	10.13	31.80	26
October	13.59	21.90	24
November	14.48	35.40	20
December	10.64	30.80	22
January	13.79	31.20	24
February	35.03	33.20	25
March	16.24	26.20	25

Source: Adur District Council Environmental Services 1999.

Notes:

SO$_2$ data are the monthly averages for the 24-hour mean.
European Community Directive sets a 'guide value' at 100–
150 µg/m^3 for the 24-hour mean.

NOx data are also monthly averages. European Community
Directive sets a 'guide value' at 28 ppb, with a 'limit value' at
40 ppb.

O$_3$ data are monthly averages based on an 8-hour average over
a day. Data monitored at neighbouring Arun District Council.

Interpretation. Data are presented for illustrative purposes
only. SO$_2$ appears to be well within the 'guide limit'. NOx is
nearer to it. Interpretation requires consideration of details of
spatial and temporal sampling to be taken into account. Daily
or hourly data could exceed limits, while the monthly average
remains well below.

"discounted" compared to those nearer' (Nicholls 2000). The idea of the
regional air-shed is beginning to catch on. New York State itself is challenging a federal regulation that would allow emissions credits (for SO$_2$ and
NOx) to be bought by power stations upwind of the state. The state wants
credits for upwind emitters to be excluded from the trading scheme and to
see them forced to make real reductions in emissions.

9.1.2 The visibility issue

Chapter 1 opened with the observation that a large part of our environmental crisis continues because people, in the richer countries at least, have
lost contact with the physical reality on which their lives depend. They may
read and hear about the issues, but they do not *see* them. This is true even
of water issues, even though that is a visible commodity and people become
sick very quickly if they drink contaminated water or if they eat contaminated fish and molluscs.

Air becomes visible when it is highly polluted; smog can look very alarming as you approach a large city from the air, or from a vantage point on an incoming highway. Even then, will any travellers turn back, or even ask themselves: 'Why am I continuing to contribute to this? Is my journey really necessary?' After all it was not so long ago that the aircraft cabin would be full of cigarette smoke, and, indeed, some people still smoke in their cars. Probably, the only people who are deeply aware of the air quality issue are asthmatics and the family members and health care workers who know them. Public ignorance may be due to the fact that many air-borne contaminants have their effects masked by other factors, or have a very long latency period, such as lead and asbestos, respectively.

Is there anything that might change this lack of visibility of the issue? As noted in Chapter 1, the climate change problem will change everything, because it is large, pervasive and affects everyone. Political pressure will mount to reduce the human contribution to atmospheric change. Other pollutants will be dealt with as a side-effect of efforts to reduce greenhouse gas emissions. How can we make the climate change issue more visible to citizens? One way is to make people more aware of their ecological footprint, and the atmospheric component of that footprint. On the web there are examples of 'CO$_2$ calculators' that make it fairly simple for people to link their energy use to greenhouse gas emissions. (See list of websites at the end of this chapter.) Perhaps the day is not too far distant when companies, government services, and even households, will examine their energy budget as critically as their financial budget, or the time available for various tasks. Already some transportation companies are offering 'carbon neutral travel' by making a contribution on the traveller's behalf to a carbon sequestration project, equivalent to the carbon release from the trip.

9.1.3 Strategies for clearing the air

Effective strategies for solving the problems caused by our uncontrolled emission of residuals to the atmosphere require the following elements:

- scientific agreement on the impacts of these residuals
- availability of technologies to reduce or eliminate these impacts (technology could be as simple as a pair of feet or a bicycle)
- public awareness that the impacts and alternatives exist
- motivation for the public to adopt the less harmful alternatives (motives may vary from strong financial incentives and regulations to altruistic green consumerism)
- strong leadership from relevant levels of government, especially, in the urban context, coming from the local level.

The process should include not only energy conservation but also reduction in carbon dioxide emissions per quantity of energy used. This opens up three immediate lines of attack:

- more efficient use of electricity
- more efficient use of gas
- fuel switching, mainly from electricity to gas (Fawcett *et al.* 2000: 2).

More detailed implications of these options are examined in sections 9.2 and 9.3.

9.1.4 Taxation

The notion of 'paying for the past' was an important theme for Chapters 5, 6 and 7. The problem is that in the field of taxation we are still very much *living* in the past too. Despite some intermittent encouragement from governments for environmentally benign technologies, the subsidies implicit in tax concessions continue to be weighted heavily in favour of fossil fuels and nuclear power. Until air quality issues and climate change become significant features of the political agenda, governments will continue to subsidise investment that produces power in the traditional ways – coal, oil, gas and nuclear power. Despite the publicity given for governmental support for green energy, most of these existing subsidies remain firmly in place. As Charles Caccia has remarked:

> Fossil fuel subsidies in the OECD energy sector are estimated to run in the neighbourhood of $30–38 billion per year. No wonder the international agenda is so slow in moving ahead and no wonder there are difficulties in meeting the commitments needed for the climate change convention. (Caccia 1996: 2)

Action responds to incentives, and the current incentives *encourage* the use of fossil fuels. In Canada subsidies to road transportation increase, while subsidies to rail are reduced. Yet 'for every tonne of freight switched from road to rail, energy use is reduced by 85 per cent' (Caccia 1996: 4). Also in Canada there are generous tax reductions for capital costs for the fossil fuel industry, up to 50% in the first year, which are not available for renewable energy, or for district heating and other fuel conservation projects. It is understandable that it is politically difficult to eliminate long-established subsidies. It is also clear that as long as such subsidies remain in place they will help to perpetuate the established pattern of behaviour, which is energy profligate. If we are to make progress on these and similar issues our governments must be prepared to tax environmental 'bads' and provide incentives for environmental 'goods'.

9.1.5 Demand management

There is an infinite variety of instruments and approaches available to manage this problem. In the context of climate change, the heart of the matter is really very simple. We know what we have to do. We have to emit fewer (preferably no) greenhouse gases. The simplest way to achieve this is by rewarding people who emit very little and charging those who emit a lot. Eventually people will get the message. Figure 9.1 sketches the outline of a scheme that would meet this objective (White 2000: 229). It might look like a taxation scheme but really it is a transfer scheme. Both within countries and between countries it transfers funds from those who emit a lot of carbon to those who emit very little ('carbon' standing for all greenhouse gases). Each country would be assessed on the basis of its average, or per capita, emission, and would then pay into, or receive from, a 'global carbon fund'. A break-even point for per capita emissions would be chosen such that the fund would be solvent, while the break-even point would be steadily lowered until the goal of climate stabilisation was reached. Within each country the national levy would also hinge around a point where the national fund would be solvent. The national ceiling would also be lowered, steadily, to contribute to the global goal of climate protection. Convergence towards an emission level which is climate-neutral gives us a real physical

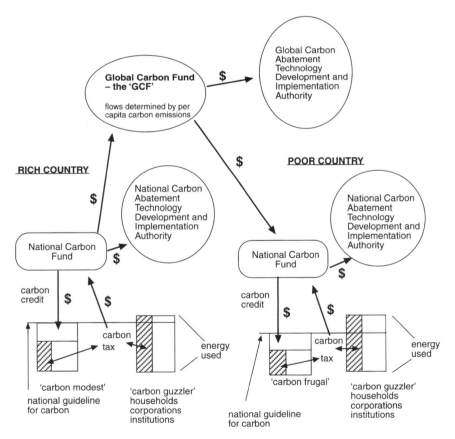

9.1 Carbon abatement, regulation and transfer scheme

endpoint to aim for, rather than small steps reluctantly taken along an endless, uncharted, and potentially fruitless path (Meyer and Cooper 2000).

Such a proposal might appear to be wildly optimistic. However, it robustly deals with important aspects of the problem that we face. First, it is just. It is based on per capita rights to the global sinks. There is no other ethical basis for allocating such rights. (These rights would have to be assigned on a national basis to a population fixed to the 1990 level, the baseline used for the Kyoto negotiations.) Table 3.2 gave some idea of the discrepancy between rich countries and poor, showing low income countries emit an average 1.5 tonnes of carbon dioxide per capita per annum, compared with 12.3 tonnes for the high income average.[1] These groupings hide the much greater range between individual countries, running from 0.7 tonnes of CO_2 for Afghanistan to 19.3 tonnes for the United States. (For details of 1990 emissions by country, see *http://www.cnie.org/pop/CO2/rankingdata.htm.*) Second, the proposal is simple and transparent. The major users of the atmospheric sink pay for that usage, and they continue to pay until excessive use has been eliminated. This approach is far simpler in conception than the schemes currently under negotiation, such as credits for carbon sinks, credits for joint implementation of emission reduction projects, and credits that can be purchased through national and international trading schemes (emissions trading is discussed in section 9.4.4). Third, it is expedient. It gives developing countries an important atmospheric responsibility *right now*, rather than leaving them in passive limbo while the world warms up and they 'develop' following the Western model. Most importantly it gives them a powerful incentive to stay 'carbon frugal' instead of imitating the West by becoming 'carbon guzzlers'. Fourth, it manages demand at the individual level, where responsibility for the problem ultimately lies. The levy would best be collected at the various points of purchase – petrol for the automobile, energy for the house, and an 'embodied energy' tax on goods and services. It would be paid for by the users – individuals, households, corporations and institutions. This would provide a strong financial incentive for the market transformation processes presented in the next section.

9.2 Energy conservation

As with the provision of water, the planning tradition in richer countries was to calculate expected demand for energy (based on extrapolations of past growth trends) and then build the infrastructure, often with public money, to meet that demand. The alternative approach – to improve efficiency to meet increased demand – has only recently been given serious attention. Otherwise energy is generally wasted until it becomes someone's responsibility to monitor it and achieve reductions.

---

[1] One tonne of carbon = 3.66 tonnes of CO_2.

9.2.1 Market transformation

At this point in world economic history it is unlikely that Western governments will revert to command and control approaches to restructuring the economy. Instead governments will have to restructure, or transform, the way that markets operate. Despite some of the excessive enthusiasm for market mechanisms they must operate within government regulations if citizens are to be protected from short term profiteering at the expense of public safety. As we have seen with the privatisation of water and power utilities in the UK, the role of the government remains substantial. In the context of climate protection, public safety requires that governments make sure that the signals they send to markets that affect energy use are consistent with government policy. Thus: 'Market transformation is a well established strategic approach, utilising a combination of policies, such as education, labels, rebates, procurement and standards, to speed up the introduction of energy efficiency technologies into the home' (Palmer and Boardman 1998: vii).

Although the central government sets many standards and regulations, there is much that local governments can do to speed the market transformation process. Once energy conservation becomes an important objective for a city, and its individual and corporate citizens, reductions can usually be made more easily than expected and with tangible savings in costs to produce a short payback period. However, the process does require a shared vision, the requisite expertise and the capital to fund the retrofit or replacement activities. It also requires incentives, as noted above.

9.2.2 Households

A household's potential for conserving energy depends on a wide range of factors including:

- awareness that problems of air quality and climate change exist
- awareness that their own behaviour is part of the problem, and hence potentially part of the solution
- awareness of technological and behavioural alternatives that could reduce their energy use
- financial capacity to adopt a lower energy lifestyle
- physical availability of certain options, such as natural gas
- access to technical information and expertise to implement a lower energy lifestyle, e.g. through appliance labelling, informed sales networks, etc
- government leadership on energy policy and the existence of a consistent set of financial instruments.

As a household's changes in energy use are a function of many small decisions taken over a long period of time, we must adopt a market transformation approach that develops a co-ordinated strategy to encourage a positive interplay between regulators, suppliers and consumers (Boardman *et al.* 1997; Palmer and Boardman 1998; Fawcett *et al.* 2000). An important element of this approach is to analyse energy use from the perspective of the 'whole house' – including space heating and cooling, water heating, appliances, lighting and personal transportation.

A significant barrier to changing energy behaviour at the household level is the low rate of turnover of the housing stock, which typically runs at 2% per annum in a Western city. This is far too slow a rate of change to contribute even to the emissions reductions schedule proposed in the first round of Kyoto reductions. Thus the required impact will come largely through retrofitting the existing stock.

9.2.3 Commercial buildings

The prospects for energy and carbon savings from commercial buildings are probably easier to realise than can be expected from households because a business probably has better access to energy use information and its impact on costs. Also the scale of the operation and the expertise available to appraise the options puts the commercial building operator in a stronger position than the householder.

The Better Buildings Partnership, a pilot scheme from the City of Toronto, has demonstrated the savings in energy and water that can be obtained from retrofitting commercial buildings. The work was financed by loans from a revolving fund set up by the city, based on expected payback periods from two to three years. There is no reason why these opportunities should not be attractive to the private sector even without governmental encouragement, as noted by Joseph Romm (see section 3.5). For example, although 'banks are a relatively clean sector', the 'potential energy savings are huge' because typically they operate many branches, and the lessons learned from one can easily be replicated in others (Jeucken and Bouma 1999: 26). Environmental audits from five German and Swiss banks and an insurance company are shown in Table 9.2. Although the figures are not directly comparable because the banks' mix of operations vary, the differences would certainly provoke a number of questions. The table suggests why 'Credit Suisse (CSG) . . . concluded that energy use is by far its most serious impact, accounting for 90% of all . . . pollution within the organisation', and also that 'UBS came to a similar conclusion' (Jeucken and Bouma 1999: 26).

In-house monitoring of environmental performance serves a much wider purpose than the immediate cost savings to be gained from improved energy efficiency and emission reduction. The exercise can begin to transform the old mindset that assumed that 'environment' had nothing to do with 'business'. Indeed, 'The increasing awareness that proactive environmental activities are an indication of good management and financial success will slowly change the old prejudice that has hitherto equated

Table 9.2 Internal environmental burden from five German and Swiss banks and an insurance company (Allianz)

Parameter	LLB	LG	BLB	Allianz	CSG	UBS
Electricity consumption (kWh/employee/year)	2886	4627	4816	4110	7500	7300
Heat consumption (kWh/m^3)	na	na	na	na	98	104
Water consumption (m^3/employee/year)	145	98	100	76	119	94
Total paper consumption (kg/employee/year)	120	113	120	258	240	252
Copier paper consumption (pages/employee/year)	3997	3895	5200	7200	8850	10600
Computer hardware (numbers/employee/year)	0.5	0.2	0.3	1.03	na	na
Business travel (km/employee/year)	321	1040	1500	3700	1700	3000
CO$_2$ emissions (kg/employee)	na	na	na	na	2850	2800

Source: Jeucken and Bouma 1999: 27.

Key: Banks: LLB – Landesbank Berlin; LG – Landesgirokasse; BLB – Bayerisches Landesbank; CSG – Credit Suisse Group; UBS – Union Bank of Switzerland.

environmental issues with costs and legislative burdens' (Hugenschmidt *et al.* 1999: 43).

Once the management of a company has developed some expertise and confidence in improving their own environmental performance they may be prepared to go further than that. They might even go so far as to embrace the view that: 'The Kyoto Protocol might convert a global environmental threat into new global market opportunities' (Hugenschmidt *et al.* 1999: 44) in a variety of business areas such as:

- corporate banking (loans to businesses for emission reduction projects)
- project finance (for Joint Implementation and the Clean Development Mechanism (JI/CDM) set up by the Kyoto Protocol)
- equity analysis and investment banking (especially regarding companies in the energy sector)
- emissions trading and brokerage (see section 9.4.4)
- carbon reduction investment funds (based on a portfolio of JI/CDM projects to reduce the risk to investors) (Hugenschmidt *et al.* 1999: 45–7).

For corporations, tackling energy efficiency through retrofitting their buildings offers all of the above advantages plus it provides a practical goal for the growing trend towards corporate environmental reporting. This is a trend that is running most strongly in Europe and, within Europe, in the north (Scott 2000: 23).

9.2.4 Industrial operations

As Joseph Romm pointed out, even greater potential savings are available from industrial operations where energy use is a significant cost component in processing materials. Manufacturing companies that have pursued energy efficiency have found greater than expected benefits, sometimes with payback periods of less than one year (Romm 1999: 9). Less dramatic success will still provide year-after-year energy reductions in seven main areas:

* energy efficiency in lighting
* heating, ventilation and air conditioning
* motors
* compressed air
* steam systems
* cogeneration of heat and power
* process redesign.

Natural gas technologies can produce power at a plant with 80 to 90% efficiency, compared with the 30% efficiency of an average fossil fuel electric power plant, which then suffers further losses in distribution (Romm 1999: 7). Even energy-intensive industries like steel, chemicals and energy production itself have made emission reduction commitments greater than those made by delegates from OECD countries for the Kyoto Protocol.

Romm believes that companies under-invest in buildings and energy efficiency because:

> managers believe that physical changes in the workplace, such as improved lighting, are irrelevant to the productivity of its workers. This error is due in large part to a powerful myth created at Western Electric's Hawthorne Works in the 1920s and 1930s. This so-called Hawthorne Effect . . . has not been supported by subsequent research. Even more shocking, the original experiments not only failed to demonstrate the effect, but actually *proved the reverse*, that work conditions can have the dominant impact on productivity. (Romm 1999: 6)

One of the simplest ways to improve the attractiveness of working conditions is to provide more natural daylight. This not only increases productivity but it directly reduces the lighting and heating bills. This has important implications for life cycle costs. As Table 9.3 indicates, 'people costs' far outweigh energy costs over the lifetime of a building which is why 'productivity savings dwarf energy savings' (Romm 1999: 78). Yet, building design is still dominated by the desire to minimise construction costs even though these are less than 2% of total costs of operation over a thirty-year period.

9.2.5 The urban system as a whole

This entire book is about understanding the urban system *as a whole*, so this section serves only to summarise a few key points. The above remarks

Table 9.3 Thirty-year life cycle costs of a building (USA)

Initial costs (including land and construction)	2%
Operation and maintenance	6%
People costs	92%
Average energy costs $1.50–$2.50 per square foot	
Average salaries exceed $200 per square foot	

Source: Romm 1999: 78.

are offered in the context of individual buildings. Much greater synergistic savings are available from individual building and production units that are linked together. Industrial parks have been set up specifically to allow one company's waste materials to be used by its neighbour as inputs to its own production. This approach is known as 'industrial ecology' because it begins to emulate natural ecological cycles on an industrial scale. Within the urban fabric, district heating and cooling systems perform a similar service, providing financial savings for the users and emission reductions for society.

Tree planting can further reduce energy demand throughout an urban system by providing shade in summer and shelter in winter. Multiple benefits can be obtained in this way:

> A shade tree planted near a city building saves ten times as much carbon dioxide as a tree planted in a forest because it reduces the energy used for air conditioning and helps to cool the city. Such tree-planting coupled with use of lighter colored roofs and road material, could *cool a city like Los Angeles by five degrees*, cutting annual air-conditioning bills by $150 million, while reducing smog by 10 percent, which is comparable to removing three-quarters of the cars on L.A.'s roads. (Romm 1999: 10)

All the approaches introduced in this section have simply looked at ways of reducing the amount of energy used to carry out a certain set of activities. Even greater impacts can be achieved by switching out of fossil fuels altogether.

9.3 Fuel switching

Energy conservation is a first step, but it is only an interim step towards an urban system that operates without recourse to fossil fuels. The argument from Chapter 6 suggested the following assumptions and options:

- Eventually we must abandon fossil fuels altogether, unless there is a technological breakthrough that allows us to sequester carbon emissions. (Biomass sequestration is too limited to absorb even our current emissions.)

- Nuclear power is not a valid alternative.
- Unused large scale hydroelectric power potential is not available in sufficient quantity to power the world.
- Technological breakthroughs *may* be made to keep the world running much as it does today, e.g. fuel cells, hydrogen power.
- However, at the present time, renewable energy is the only functioning alternative.

The first steps in fuel switching will follow the fossil fuel chain from coal to oil to gas. This is largely how Britain has achieved its early emission reductions, following a path that it would have followed anyway as gas produces not only cleaner electricity than coal, but also produces it more cheaply. This interim step is an important one to start the process of moving towards a low carbon economy. The more difficult step is the next one – moving off fossil fuels and on to renewable energy.

9.3.1 Renewable energy
The principal sources of renewable energy are:

- passive solar
- photovoltaic solar
- wind power
- biomass
- tidal
- geothermal (marine and terrestrial)
- deep lake water cooling.

The reason why these energy sources are not more widely used is that they are rarely price competitive with conventional fossil and nuclear fuels. They are close but they cannot compete against the subsidies the conventional fuels receive in the form of tax breaks for exploration and capital costs, plus the fact that the externalities involved in the use of fossil fuels and nuclear power are absorbed by other sectors of the economy. Ironically, sources like photovoltaic power would rapidly become price competitive if they had sufficient market share to bring the unit price of supply down.

If the existing subsidies were removed the market would reshape itself to allow renewable energy in slowly. If a carbon tax were imposed – and if the present subsidies were removed – then this shift would be much more rapid. Energy companies like Shell and BP are acting on the assumption that global warming will force governments to impose carbon taxes in order to reduce emissions of greenhouse gases. They are investing now in renewable energy.

However, the longer governments prevaricate the more dangerous and difficult the transition to renewables becomes. As temperatures rise there will be a significant rise in the demand for air conditioning, especially in Europe where households and older institutional buildings have not yet installed it.

A surge in cooling demand would have serious consequences for fossil fuel use as, so far, air conditioners are powered from the electricity grid which continues to rely on coal and oil in many countries. This could create a dangerous positive feedback for global warming. Given that we are already committed to a certain amount of warming it is essential that renewables quickly provide a significant input to the grid, or else stand-alone, solar-powered cooling units should be developed for households and offices.

9.4 The transportation challenge

We are seriously failing in the transportation sector. We produce cleaner fuels, but we make more mileage. Trips are getting longer, are being used increasingly for non-work purposes (such as recreation), and there is more energy being wasted by congestion. Congestion is no longer a city centre problem but a city-circle problem – from the 401 (Toronto), to the Periphérique (Paris), to the M25 (London). This is an urban problem to which we must find an urban solution. Of all the traps we are leaving out there for developing countries this is by far the most serious, because the automobile is the ultimate rite of passage from poverty to material wealth.

9.4.1 Citizens' attitudes and political will

Householders may simply be ignorant of the environmental options in the house, but drivers can be fiercely defensive about their automobile lifestyle. Protests against fuel price hikes are delivered with almost religious conviction. Surveys report that road users strongly believe that only a small percentage of what they pay in road tax and fuel tax is reinvested in road infrastructure through new construction and maintenance. They believe that more routes will relieve congestion despite decades of experience that show that any improvement is small and temporary, as faster journey times attract more users, such that the additional capacity is rapidly swamped with additional use. This is true for both intra-urban and inter-urban trips.

> Whatever road construction policy was followed, the amount of traffic per unit of road would increase, not reduce. . . . this implied that all available road construction policies only differed in the speed at which congestion would get worse. Therefore demand management would force itself to centre stage as the essential feature of future transport strategy, independently of ideology or political stance. As a

result . . . it was possible for consensus to develop, not derived from political or ideological agreement, but derived from the technical impossibility of keeping pace with traffic growth. (Goodwin 2000: 15)

Road users also neglect to factor in the externalities of road transport which include:

- accidents, including large numbers of deaths
- damage to roads
- congestion costs
- noise
- air pollution
- global warming.

Even with very modest costs attributed to global warming Maddison *et al.* calculated costs that are approximately three times the revenue from road taxes (Table 3.4).

Perhaps it is not surprising that few politicians have mustered any political will to tackle this serious problem. On both sides of the Atlantic politicians are desperately anxious to proclaim that they are not 'anti-car'. Thus it is a great challenge to know how to change behaviour in this sector. It is likely that the zealots will never be convinced. Change may come slowly as some members of society adopt other transportation options and as the cost of driving cars running on fossil fuels mounts, once carbon taxes become widespread.

9.4.2 Alternatives to fossil fuel-powered automobile use

It is possible that some alternative fuel such as hydrogen, batteries or fuel cells will be developed to power automobiles and that they will continue to proliferate around the world. Certainly politicians will breathe a sigh of relief if this scenario unfolds. However, in some ways, it will be a shame if it does because automobiles have other disadvantages apart from emitting pollutants to the air. The space they require means that they will always cause congestion and require a great deal of urban land, while creating noise and causing accidents. We need a vision of a car-free world.

With a supportive land use pattern that included variety and density, walking would have the least environmental impact, followed by cycling. A new town like Poundbury is designed to encourage travel by foot and bicycle. Other communities are struggling to redesign their transportation lanes to make this possible too. All too often the efforts are focused on one option – such as cycle paths – whereas our vision must be integrated. In the transportation sector that means that we must think 'multi-modal'. Results so far are very variable: 'More than a third of Danish and Dutch rail passengers pedal to the station. In Germany 15 percent of rail travellers take their bikes. But in Britain the number is less than 1 percent' (*The*

Independent on Sunday, 9 July 2000: 12). The reasons for Britain's lack of success in this area are very obvious:

- Many railway stations are accessible to high volume car traffic but difficult for bicycles to access.
- There is a lack of storage for bikes at stations.
- It is very difficult to take a bike on the train as there are few places and often they must be booked in advance.

This kind of 'planning' is definitely *not* multi-modal.

Similarly there has been very little planning to deal with the trend towards driving children to school, and this might involve a two-way trip twice a day if the school drop-off is not part of another commuting journey. Of the private vehicles on the road approximately 80% of them carry only the driver.

There are a host of actions that can be taken to reverse this situation, with initiatives ranging from the neighbourhood to the municipality. In some communities parents have organised what they call the 'walking bus', whereby they take it in turns to escort groups of children to school. Some companies encourage car-pooling and offer bus passes, instead of providing free parking for every worker's individual car.

On a larger scale, cities like Oxford have developed a strategy over the years to reduce traffic in town (Townsend-Coles 2000). This included:

- routeing through-traffic via a ring road
- building 'park-and-ride' facilities at each of the main entry points into town for incoming workers and visitors
- providing rapid public transport, including priority lanes for buses
- constructing cycle paths
- pedestrianising major streets
- increasing the price of parking in town.

9.4.3 Incentives and barriers

Alternatives to an intensive fossil fuel-based urban society do exist. They are not hypothetical, nor need they be expensive or repressive. The first thing to be done is to remove disincentives such as perverse subsidies to fossil and nuclear fuels. The next priority is to redesign unhelpful land use patterns that make it difficult for people to live a low carbon lifestyle. Lastly there must be some positive incentives for people to look at alternatives to automobile dependence. For example, if people want to save money by cycling there should be supportive infrastructure to help them to do so. Compared with freeways and multi-storey parking lots this low carbon infrastructure does not need to be either expensive for the public purse or intrusive on the landscape. It could be as simple as a cycle lane free from heavy traffic and a secure place to store a bicycle at the destination.

9.4.4 Emissions trading

There are many schemes afoot to 'clear the air', especially now that climate change is being recognised for the serious threat that it is. There is concern that reducing greenhouse gas emissions will entail both the loss of a much-loved, energy-intensive lifestyle and – paradoxically – the loss of international competitiveness for those countries that take the first steps. These fears explain why the trading of emissions is attractive to those parties that fear the changes the most. The economic justification is that the most inefficient energy users should reduce their emissions first because that would be most cost-effective. This book is not the place to delve into the pros and cons of the various proposals as this has been done elsewhere.

The point to be raised here is related to the argument that has been made concerning the opportunities that cities have to reduce emissions. Cities today – from an energy perspective – are hugely inefficient. If trading is to take place then municipalities should be allowed to participate. This would give them another incentive to reduce emissions and would provide them with some of the finance that the initiatives require. Some of the international implications of carbon trading are discussed in Chapter 11.

9.5 Conclusion

There are technologies available now that could be introduced to clear the urban air. We continue on our present path because we are in the grip of inertia. Most citizens are unaware of what their machines and buildings emit, and thus what they inhale to the detriment of their health, and unload into overloaded sinks to the detriment of the planetary niche on which they depend for their existence.

Outdated assumptions clog the thinking process so that intelligent choices are not made. Society is still actively subsidising and advertising the very technologies that will undermine the urban lifestyle they currently enjoy. The first remedial step is to monitor our impacts on the air and to measure the full costs of our current technological preferences.

Part of the problem lies in the implicit assumption that we are happier if we accumulate more material goods and if life becomes increasingly complex. There is an unwillingness even to consider the idea that part of the solution to our problems might be as simple as a bicycle or our own feet. Like a fractious child clutching a toy we cannot consider just letting go.

9.6 Websites

1. *Atmospheric Change in the Toronto – Niagara Region* can be downloaded from: *http://www1.tor.ec.gc.ca/earg/tnrs.htm*

2. British Petroleum re-branded as 'beyond petroleum': *http://www.bp.com*

3. Calculate an individual's greenhouse gas emissions: *http://www.climcalc.net/eng/Intro_1.html*

4. Converting a power company to solar power: *http://www.utoronto.ca/envstudy/solar.htm*

5. Corporate environmental reporting: *http://www.corporate-register.com*

6. International carbon emissions 1990: *http://www.cnie.org/pop/CO2/rankingdata.htm.*

7. US government site for procuring low energy design and consulting services: *http://www.eren.doe.gov/femp/techassist/low_energy.html*

8. World Energy Council: Energy Data Centre: *http://www.worldenergy.org*

9.7 Further reading

Boyle, G ed (1996). *Renewable Energy: Power for a Sustainable Future*. Milton Keynes, Oxford University Press in association with the Open University.

Brower, M (1993). *Cool Energy: Renewable Solutions to Environmental Problems*. Cambridge, MIT Press.

Elliott, D (1997). *Energy, Society and Environment: Technology for a Sustainable Future*. London, Routledge.

Fawcett, T, Lane, K and Boardman, B with other contributors (2000). *Lower Carbon Futures for European Households*. Oxford, University of Oxford, Environmental Change Institute.

Grubb, M (1995). *Renewable Energy Strategies for Europe. Volume 1 – Foundations and Context*. London, Royal Institute of International Affairs and Earthscan Publications.

Maddison, D, Pearce, D, Johansson, O, Calthrop, E, Litman, T and Verhoef, E (1996). *The True Cost of Road Transport*. Blueprint No 5. London, Earthscan.

Nicholls, M (2000). 'Ontario looks across borders', *Environmental Finance*, April, 6.

Romm, J J (1999). *Cool Companies: How the Best Companies Boost Profits and Productivity by Cutting Greenhouse Gas Emissions*. London, Earthscan.

10

Water – our most precious resource

The quality of water determines the quality of life. (Todd 1996: 42)

The prospect of water scarcity is very real with implications for regional peace, global food security, the growth of cities, and the location of industries. . . . Urbanization will enlarge the claims on available supplies because of higher per capita water consumption in urban areas. . . . The ability to supply safe, clean water and adequate sanitation, already stretched, will be severely tested. . . . Agriculture is already competing for available water resources with urban and industrial uses and competition will only intensify with time. (World Bank 2000: 29)

10.1 Integrated watershed planning

The pressure is on to manage water resources in an integrated fashion. This implies far more than recognising the need to manage entire river basins as a hydrological unit. The basins themselves must now be managed as an integral part of the hydrological cycle, which is itself about to undergo significant modification in response to the changing climate. To this must be added the demographic and political context. Most of the global population increase will happen in developing countries, which are already the most water-stressed regions. However, this does not mean that the rich can afford to be complacent about this issue, because their own water supplies are poorly managed and under stress. Furthermore, water scarcity in the poorer countries will add to the pressures that generate flows of environmental refugees towards the richer countries.

The stock of freshwater is the planet's most valuable resource for human civilisation. Unfortunately, in the rich countries, we have grown up taking its availability for granted. Many people assume that water supply is an environmental service for which there is no charge. Water treatment and reuse came into being to protect human health, rather than to conserve water. Until recently water quality was the key criterion for water management, while losses through leaking pipes and evaporation were regarded as a secondary matter, or even completely ignored.

We can blame the inefficiency of agriculture, as the major user, or industry, as the next biggest consumer, but the managers of cities and their citizens owe it to themselves to safeguard the water supply. We can live without automobiles; we can live without electricity; but we cannot live without clean water.

As with all urban environmental issues, the complexity of the problem may appear daunting. Any improvements involve a host of technological, economic, medical and ethical considerations. Even so, the problem of urban water management can be summarised very simply: sometimes we have too much, sometimes too little, and we must always safeguard against contamination. As the introductory quotation from the World Bank suggests, we are now entering an era of global competition for water.

10.1.1 Living within our hydrological means

It is possible that a technical fix will be found, such as a cheap technology for desalination that also solves the residuals problem, especially the salt. Just as we might solve the carbon dioxide problem by sequestering it underground. However, these scenarios are unlikely to appear, and certainly not without some unintended side-effects. The lesson we have yet to learn is to live within our biospheric means. For water that means learning to live within the natural hydrological cycle without impairing the supply of water that we need. For people in the rich countries the amount we 'need' must then be reduced to the amount that is naturally available. Just as we need to reduce our carbon dioxide emissions to half a tonne of carbon per person per year, so we need to reduce our daily, per person, residential water use to about 200 litres at current population levels.

The organisational structure that governs the planning and management of water resources varies greatly from country to country. Some bodies are public, some private, others are in transition from one state to another. Extraction from some rivers is governed by international treaty. As demand rises everywhere there is pressure to revise treaties and other agreements. Without the kind of technical breakthrough mentioned above, or some fortuitous (and unexpected) side-effect of climate change, we have now come to the point where we have to accept that the supply of raw water cannot be expanded much further. Indeed, some of our existing 'extensions' have led to the widespread salinisation of irrigated fields and other long term contamination problems. Like the overloading of the atmospheric sink with greenhouse gases, we have now entered a new phase where the old assumption about endless new supplies has come to an abrupt end.

Modern Western cities have evolved to meet needs that are quite unrelated to the key concern of this chapter, which is the management of water quantity and quality to ensure the availability of scarce freshwater resources. That is really not surprising as water was rarely seen as an important constraint to urban design even in regions with low rainfall and limited surface water resources. Until now the growth of cities has been driven by economics, demographics and politics, not biospheric issues like water resources and carbon dioxide sinks. Cities that continue along this historic path, blinkered from the environmental reality on which they depend, will significantly enhance their exposure to risks such as flooding, drought and water contamination.

This is a common problem for cities as politically dissimilar as Los Angeles and Beijing. Both have been constructed in semi-arid regions and now contain approximately 12 million inhabitants, being among the 10 largest cities in the entire world. Each struggles to exist in physical conditions for which their numbers are totally unsuited. There has never been any suggestion that either city had reached some physical limit to growth, or that the price of water should rise to a point where further expansion becomes uneconomic. For each it was assumed by the powers-that-be that such a trifling thing as the unavailability of water should not hinder growth. For Los Angeles the solution was to pump water down from the north, and from the Colorado River to the east; nor was this demand limited to what these faraway sources could sustainably supply. Beijing likewise cannot meet water demand from sources in north China anymore; instead it plans to undertake the breathtaking South–North water transfer from the Yangtze River. Contrived economic benefits and political pressure drive such decisions, not ecological, social or financial coherence.

10.1.2 The new reality

From now on, water management has to deal with two new factors. The first factor is that demand is running into the limits of the supply of freshwater – under the current hydrological regime. The second factor is that climate change will greatly increase the variability underlying that supply. This implies that we have to meet both challenges simultaneously.

To deal with the first point we have to reverse our entrenched hydrological habits, which are incredibly thoughtless and wasteful. We have to learn to value water for what it is – indispensable. As the constantly rising demand for water runs up against the fixed supply we will have to abandon two of our most cherished assumptions – that we can over-pump groundwater beyond the recharge rate, and that we can always import water from greater distances. In general we have to be aware of the dangerous point to which our ignorance has brought us; we must also abandon our technological complacency which is based on end-of-pipe response to emergencies. It is true that more water can be provided by desalination, and it is true that *Cryptosporidium* can be screened out by reverse osmosis; but technologies such as these deliver water at a much higher price than we have paid up till now.

In order to deal with the second point, we have to learn to manage our freshwater resources much more carefully, in the expectation that many rainfall regimes will produce more dry spells and more intense rainfall events. The increased intensity of these events will also pose increased risk of contamination. In other words, climate change threatens urban water management on all three fronts – too much water, too little water and contamination.

10.2 Planning for climate change

10.2.1 Greater variability in water availability

There can be little doubt that the changed precipitation regime predicted for climate change is already with us (NOAA/USDA 2000). The intensification of the hydrological cycle has accentuated seasonality and increased the intensity of precipitation events. As a matter of self-preservation urban governments should develop water budgets on the assumption that this is the type of precipitation regime with which they will have to learn to live. Obviously this can be done only with the active co-operation of all the water users in the basin.

First, they need to know how much raw water will be available to them from 'their' river basin. Which basin is 'their basin' is in itself a vexed question – just as it was difficult to decide how a city should take care of 'its own' solid waste or which gaseous emissions fell under which jurisdiction. Just how this 'boundary problem' will be resolved is not clear at this time. It will require either strong central direction or novel approaches to bottom-up co-operation.

Water basin transfers should be regarded as a thing of the past, because all supplies will become tight in the near future. Even huge water bodies like the Great Lakes will be stressed as lake levels fall (Mortsch and Mills 1996). If cities adopt this limitation it will immediately put the emphasis on recycling and safeguarding the water from avoidable contamination.

The same stricture applies to groundwater. Before industrialisation, groundwater was the preferred source of supply because it could be assumed to be clean. It was also a more assured source of supply than surface water because it was less directly dependent on recent rainfall; nor was it subject to evaporation losses. Its assumed purity was reflected in surprise that groundwater used in north London could be contaminated by *Cryptosporidium* (as reported in Chapter 7). Thanks to industrialisation – including the industrialisation of agriculture – this assurance of quality for groundwater has been lost. Groundwater – the champagne of 'our most precious resource' – has been routinely over-pumped and contaminated. As a first priority we need to restore the reliability of groundwater, as this should be our first line of defence against the greater variability of precipitation that we expect to come with climate change.

More storage capacity will also be required, both to hold excess supply from heavy rainfall events, and to provide water capacity for dry periods. Storage capacity in this context not only means reservoir capacity but also groundwater management. The ability of a water management regime to withstand greater variability of rainfall can be tested through simulation modelling. For example, a model was developed for the Grand River basin, in southern Ontario, to predict the impact on it of the record rainfall that fell on the Saguenay River basin, in Quebec, in August 1996, causing severe flooding and loss of life. The model demonstrated that the management system in the Grand River basin could withstand even that amount of rainfall without the integrity of the system being breached (Southam *et al.* 1999).

In addition to simulation modelling planners may also compare recent experience with their historical experience. However, the record of recent change is short, while scientists and engineers tend to be cautious about anticipating unprecedented developments like climate change. The following is a fairly typical official assessment of a disquieting trend which comes from the UK Institute of Hydrology with reference to the 1995/96 drought:

> Notwithstanding the recent notable drought episodes, rainfall over Britain for the 1990s thus far is very close to the long term average. The severe stress on our water resources and river systems is attributable to the *very atypical* [italics added] spatial and temporal distribution of the rainfall – exacerbated by the *very high* evaporation losses resulting from the *exceptionally* mild conditions. Taken together, the last 25 years have seen an accentuation of the north-west/south-east rainfall gradient across the British Isles and a tendency towards a more distinct partitioning of rainfall between the winter and summer half years. Importantly, the recent past has seen a cluster of hot, dry spring/summers *for which there is no close modern parallel.* This clustering raises the possibility that the level of risk adopted by the water industry . . . may no longer be appropriate. (UK Institute of Hydrology 1995/6: 17)

The report concludes cautiously:

> Fortunately, careful documentation and analysis of current hydrological conditions and their impact on water resources and the aquatic environment are providing valuable insights into the type of problems *likely to be encountered in a warmer world* [italics added]; these insights will help shape the development of more effective water management strategies *should hitherto rare weather conditions become rather more familiar in the future.* (UK Institute of Hydrology 1995/6: 18)

The phrases in italics strongly suggest that something very unusual (and potentially threatening) is being experienced, while the overall tone of the judgement is imbued with reassurance that everything is, nonetheless, under control!

To this managerial caution should be added the fact that budgetary pressure on public expenditure, over several decades, has produced a minimalist solution to urban infrastructure. Facilities such as reservoirs are made no larger than necessary to accommodate expected variability. Storm drains, formerly built to handle the ten-year flood return period, have been put in to handle the five-year flood, or sometimes even the two-year flood. This margin of error could prove disastrously small under climate change scenarios, as it is a move in entirely the wrong direction. In so far as the historical record can be used at all in the infrastructure design process, planners must be prepared to cope with the hundred-year flood.

10.2.2 Changing shorelines

A rise in sea level is the most predictable consequence of global warming. It will have profound implications for managing water resources in all coastal and tidal communities. The scale and cost of adjustment will depend on the topographic detail of each site. New pipes will need to be laid and new pumps installed. As the sea level will continue to rise for decades – whatever mitigative measures are undertaken for greenhouse gas emissions – the new drainage will need to be flexible and cheap to modify. It is fortunate that the days of discharging untreated sewage to the ocean are coming to an end anyway, at least in the Western world. Otherwise the health implications of a rising sea level would be even more serious.

Initial response to predictions of a sea level rise was in terms of the seawalls that would have to be built. More considered reflection has shown that walls will not solve the problem, except in the short term and at very high cost. Local authorities along the coast of the British Isles (especially southern England where the land is sinking) are now contemplating a 'managed retreat' of the shoreline, by reconstructing wetlands to provide a buffer zone.

For those cities like Venice and Bangkok, which are sinking because their groundwater is being over-pumped, the need to redress their hydrological balance is doubly urgent. It is pointless to build sea-wall defences if the root cause of the problem remains unaddressed.

Cities on the Great Lakes, and other large freshwater bodies in continental interiors, will probably have to deal with the opposite problem of falling water levels, due to increased evaporation. This may require the extension of freshwater intake pipes; it may also expose new shorelines that are highly contaminated by shipping and industrial discharges.

10.3 Facing the urban flood issue

The flood risk discussed in the previous section is the 'river basin flood risk' originating in precipitation events, or flooding due to snow-melt in the spring. Cities can protect themselves from this risk only in co-operation with all the other water managers and water users in the basin. However, other contributing factors to the flood risk are entirely within the responsibility of urban authorities. These relate to land use planning as it affects both water quality and flood control.

The redesign of an urban hydrological system begins with restoring the urban system to play an active, positive role in the natural hydrological cycle. This implies an increase in the porous surface area so that a larger percentage of rainfall goes into the ground, or is retained in ponds, rather than becoming runoff to surface water, picking up contaminants from road surfaces along the way. It means separating the combined sewage and stormwater systems as these present a serious risk of contamination in times of heavy flow. The increase in porous land use can be provided by naturalising parts of the urban landscape and requiring developers to include 'green infrastructure' as an integral part of their plans. The increase in green space will bring the additional opportunity to extend the urban forest, which in turn will reduce energy demand for cooling in the summer. These land use changes need to be supported by a more rigorous system for controlling the introduction of contaminants into the hydrological system, from industry, agriculture and households. It is also helpful to identify streams that have been covered over and restore them to the surface, through a process known as 'daylighting'. This may encourage citizens to remember that they too are an integral part of the hydrological cycle.

Changes on this scale will require input from a broad partnership. This may sound idealistic, but the fact is that the current mismanagement of our urban water resources is costly for everyone. Leaking pipes benefit no one. Nor do urban floods that damage households and leave the victim unsuccessfully claiming compensation from the municipality and the insurers, neither of which consider themselves responsible (Kulkarni 2000).

The unresolved problem of liability for urban floods (discussed in Chapter 7) is similar to the liability barrier to cleaning up contaminated land. Clearly both involve a combination of public and private responsibility and the now familiar problem of today's citizens and companies being obliged to pay for problems created in the past. An important difference is that we are knowingly increasing the potential risk of urban floods by permitting construction to go ahead in flood-prone areas and localities adjacent to them. We need to move away from a planning procedure that considers site applications in isolation; instead, each application should be considered in terms of its role in the urban hydrological cycle. For cities that are already largely 'built out', steps should be undertaken to increase the porous area to no less than 20% of the total, or a similar figure appropriate to local hydrogeological conditions. The need to reduce the urban flood risk is just one additional reason for restoring the biological productivity of the urban area by introducing green infrastructure and extending the urban forest.

10.4 Water supply and energy use

Most cities require some electrical energy for pumping water and monitoring stocks in the system. Indeed:

> Across the United States, energy consumption accounts for 50% to 75% of the cost of operating municipal water systems. Of this, pumping the water often consumes 80% or more of the electricity used in water distribution and treatment and therefore represents a prime opportunity to save energy and dollars. (US Department of Energy 1994)

Under climate change, water supply systems need to be more closely managed to deal with the expected increase in variability of precipitation. Following the adage that 'what you do not measure you cannot manage', water suppliers will need better real-time information of water availability and flood control capacity. Computer control systems such as supervisory control and data acquisition (SCADA) systems have been developed to reduce energy use and to monitor the functioning of water supply and treatment. Significant cost savings produce payback periods from two to five years, similar to the energy savings reported from manufacturers in the previous chapter. Developments such as these are all part of the evolution of the 'intelligent city' through which urban systems will develop a capacity to monitor their functions continually and restore malfunctioning parts (White 1994: 161–4).

For example, the City of Fresno, California,

> pays about $5 million per year in water distribution power bills, about $725 000 less than it would be paying without the SCADA system. Completed in 1988 at a cost of $3.2 million, Fresno's system is based on a VAX computer system that communicates by radio with the 200 wells that supply the city's water. (US Department of Energy 1994: 1)

The system produces the savings by:

- selecting which wells should be pumping water, based on the energy required to pump each gallon
- taking into account the time-of-day electrical rates for each well
- controlling well line pressure, to make sure that pressure is kept no higher than necessary, which saves both energy and water that was lost to leakage
- chips installed on each pump warning of mechanical problems and leakages which allows cost-effective maintenance to be carried out.

Results from the United States suggest that this type of real-time monitoring and control of water supply will reduce operating costs by 10 to 30% per year, as well as reducing water wastage and increasing customer satisfaction. This approach helps to reduce operating uncertainty, which is a move in the right direction given that we can expect climate change to introduce new uncertainties. As the historical record of precipitation, stream flow and water availability becomes less reliable it is even more important that we have a better knowledge of the current operating status of all aspects of the water supply system.

There are other neglected opportunities to be found in the interplay between hydrological and energy issues. For example, a great deal of water is used purely to cool power stations and other industrial operations, and then voided to a nearby water body, often with harmful environmental impacts. This warm water is a resource that should be used to heat space and water in buildings. Yet this is rarely done. Similarly there has been a proposal on the drawing board for several years to use deep lake water to cool buildings in downtown Toronto (Canadian Urban Institute 1993). Any city adjacent to a cold body of water could develop such a system. As noted earlier, modern (industrialised) human beings are not very efficient in thermal terms.

10.5 Water treatment

Since the end of the nineteenth century Western cities have relied heavily on chlorine to disinfect their water. Sewage sludge would then be dewatered and the residue could be burned or spread as fertiliser. Problems can occur when contaminants enter the waste stream, making it dangerous to burn the residues or to spread them as fertiliser. Also new bacterial threats like *Cryptosporidium* can overwhelm a conventional treatment system following heavy rainfall events.

These problems, plus the need to conserve water supplies, have led pioneers like John Todd to consider other options such as the Living Machine, which cleans sewage and industrially contaminated water using natural biological processes, mostly consisting of higher plants, such as reeds, through which the influent flows (Todd 1999: 136). The effluent is tested using the standard indicators such as biological and chemical oxygen demand, total nitrogen, and total suspended solids. Living Machines are

> a family of ecological technologies that, like most conventional machines, are designed to undertake directed work . . . to generate fuels, grow foods, convert wastes, purify waters, restore degraded environments, and regulate and improve the quality of climate within buildings. (Todd 1999: 132)

These installations now operate in eight countries and 15 American states, the largest being a Brazilian plant that treats the equivalent of over a million gallons of sewage per day (Todd 1999: 133). A town in Germany with 100 000 people is served for sewage treatment only by a reed bed. In Ontario and Quebec they are used instead of the traditional septic tank, requiring a standard allocation of 4.5 square metres of reed bed per person, and costing only one quarter the price for construction. Living Machines can be designed to produce zero-discharge. Thus, in arid climates they could make an important contribution to water conservation. These systems have been tested in a variety of climatic regimes including low temperatures.

However, it will take time for the biological remediation of wastewater – using the Living Machine approach – to be considered as a viable alternative to conventional water treatment (Tang 1996). Like energy conservation ('turn out the light, put a lid on the pan'), composting, cycling and walking, it is perceived as a 'low tech' approach and therefore unlikely to gain acceptance easily. Like the other components of ecological urban life, it has to run against the trend of the past 100 years.

In general, water treatment strategies have been designed to react to whatever is likely to be put into the waste stream, without urban managers having much control over what that might include. Instead they are left to devise end-of-pipe solutions. It would be better to establish land use patterns and regulations that reduce the chance of contaminants entering the waste stream. For example, industries and housing could be designed as an integral part of Living Machine technology, to treat most, if not all, of their contaminants without having to pump liquid waste through a series of pipes to a sewage treatment plant, which then must transport its dry waste to a landfill. As John Todd remarks, 'it is our hope that Living Machines will start a trend toward treating wastes, including sewage, close to their points of origin' (1999: 133). The use of the term 'close' is like recommending that communities landfill 'their own wastes' or that local authorities manage air emissions in their jurisdiction. Action requires clarification of the boundary issue.

As in the case of contaminated site remediation, a biological solution is available to meet the sewage treatment challenge. What the Living Machine offers is a closed circle alternative to our unsustainable chemical throughput. The fact that mechanical and chemical treatment systems were the first large scale systems developed in the industrial age does not mean that we have to rely upon them forever. Even so, any treatment technology is an end-of-pipe solution. Our primary concern should be an understanding of the inputs followed by a reduction of problematic inputs wherever possible.

Reaction to the *Cryptosporidium* threat (described in Chapter 7) illustrates the difficulty that modern technology faces when expected to deal with inputs that it cannot control. Allowable levels of chlorine will not eliminate all the oocysts, as they are extremely hardy. The only successful screen is reverse osmosis, which is very expensive because it requires a lot of energy to force the water through a plastic membrane (Davis and Cornwell 1998: 771–2). Yet it is believed that the contamination originates in animal wastes and human sewage which can be prevented from reaching the water supply by adopting appropriate waste handling and land use practices, especially on farms. What is not known is why outbreaks are becoming more common, but it could simply be that climate change is responsible for more high intensity rainfall events around the world, and that these events rapidly flush the waste into the streams and groundwater that feed into our water supply. If that is the case only a preventive, land use approach will be able to resolve the problem.

The need for this kind of pro-active response is a further reason for closer co-operation between urban authorities and the rural catchment area. Some authorities, such as New York City, have undertaken exactly that kind of approach towards farmers in their catchment areas in upstate New York. This is simply a recognition of the fact that farms – inadvertently – are suppliers of water, as well as food, to cities.

10.6 Demand management

Like the automobile, clean water, available on tap at low cost, started out as a great potential benefit to society. However, as its use has become ubiquitous in modern cities it is being used very carelessly and it is costing a great deal to treat the wastes. Most modern cities sited on a lake, estuary or ocean have so polluted the surrounding beaches that they can be used for neither fishing nor recreation. It might seem that there must be a trade-off between reducing water demand and public health, but this is not the case at current levels of use in Western cities. It is simply that water should no longer be regarded as a (nearly) free good. People should enjoy a right to water, but not the right to waste it.

The level of personal water consumption can be scaled down, safely, to 200 litres per person per day. This can be achieved by a combination of metering water supply and charging per litre used, and by supplying improved technology, such as low-flush toilets and efficient showerheads at little or no cost. For example, for Toronto, it has been proposed that the city retrofits 13-litre toilets with toilets that use only 6 litres per flush (Loughborough 1999: 125). Metering alone will reduce household water use by 20%.

Water is like solid waste in that households typically have no idea how much they use, and little idea how much they pay for it. If both the quantity used and quantity paid became more visible, then people are more likely to appreciate the value of a reliable water supply. This is even more true for businesses and local authorities themselves for whom the bottom-line savings can be considerable. For industry and agriculture many countries seem to encourage waste by offering bulk rates for large users, or, in the case of agriculture, charging a flat rate for irrigation water. Even in poor countries it has been shown that under-pricing resources only serves to mis-allocate them, rather than encourage their efficient use.

10.7 Conclusion

As in the case for energy use, with water we are coming to the end of an era. Whatever justifications might have been offered in the past, we can no longer afford to under-price and waste our most valuable resource. Cities will be the first to suffer if water shortages become more common, because, without adequate water supply and treatment, cities become a target for a variety of water-related disease vectors. However, there is a big difference between the 200 litres per day for a person's household needs and the 600 litres that some urban dwellers use daily. The disconnection of the city from the natural hydrological cycle is symptomatic of the need for change. Urban water users – households, government, commerce and industry – could all use water more efficiently without any loss in amenity and productivity, by targeting investments with short payback periods.

Accommodation must be made with agricultural users where they are competing for the same water source. In most cases their water usage will be even less efficient than users in the city.

Above all we have to reconsider a pattern of economic growth that has totally ignored the realities of the hydrological cycle. We cannot encourage the growth of cities in semi-arid areas, knowing that their growth can be supported only by importing vast quantities of water over great distances, drawing supplies from areas that are already, or will soon be, in need themselves. Inter-basin transfers – like over-pumping groundwater – must be consigned to the past.

10.8 Websites

1. Environment Canada, Adaptation and Impacts Research Group, Climate Change and Variability in the Great Lakes: *http://www.msc-smc.ec.gc.ca/airg/glslb/pubs/sympMay97.pdf*

2. UK National Water Archive Office, Institute of Hydrology: *http://www.nwl.ac.uk/~nrfdata/front_nw.html*

3. UK Natural Environment Research Council, Centre for Ecology and Hydrology: *http://www.nwl.ac.uk/ih/*

   see, for example, a very useful list of hydrology hyperlinks at: *http://www.nwl.ac.uk/ih/devel/wmo/*

   see also hyperlinks for research institutes in Hydrology at: *http://www.nwl.ac.uk/ih/devel/wmo/hhcres.html*

4. United States Department of Energy – urban water system energy costs: *http://www.eren.doe.gov/cities_counties/watersy.html*

5. Water supply and sanitation at the World Bank: *http://www.worldbank.org/html/fpd/water/topics/tech_wtreat.html*

10.9 Further reading

Davis, M L and Cornwell, D A (1998). *Introduction to Environmental Engineering* (third edition). Boston, McGraw-Hill.

Ferguson, B K (1998). *Introduction to Stormwater: Concept, Purpose and Design.* New York, Wiley.

Henry, J G and Heinke, G W (1989). *Environmental Science and Engineering.* Englewood Cliffs, Prentice Hall.

Timmerman, P and White, R R (1997). 'Megahydropolis: coastal cities in the context of global environmental change', *Global Environmental Change*, 7, 3: 205–34.

Todd, J (1999). 'Ecological design, Living Machines, and the purification of waters', in *Reshaping the Built Environment: Ecology, Ethics and Economics.* Ed C J Kibert. Washington DC, Island Press: 131–50.

Part V

Conclusions

11

International issues

As the issue of climate change began to develop from a purely scientific matter into a political issue in the late 1980s, countries began assessing their interests, objectives and concerns, and the steps that should be taken to develop an international response. . . . Cooperation and action to limit climate change is complex because serious responses could reach deep into countries' economic and political interests. (Grubb et al. 1999: 27)

There's no way in the world we'll be able to convince our friends in India or China, which over the next 30 years will become bigger emitters of greenhouse gases than we are, that they can take a different path to development, and that we're not trying to keep them poor, unless we can demonstrate that we have . . . evidence that a different way will work. (US President Bill Clinton to the Democratic Leadership Council, 28 May 2000)

11.1 International aspects of human impacts on the biosphere

The awkward phrase 'rich and poor countries' has been used throughout this book, with a brief acknowledgement of the complexity that it glosses over. The dichotomy still has some meaning, but less and less by the month, as globalisation of the economy becomes a reality. All national governments are losing power relative to the global reach of corporations. The abandonment of communism, in all but name, has created a whole set of 'emerging economies' or 'economies in transition', emerging into an unpredictable world, and in transition to we know not what. China and Vietnam have even managed to take this step into the unknown, while the communist party remains in power. Meanwhile Central Europe is looking more like Western Europe every day. Except in Africa, and a handful of countries in Asia and Latin America, the logjam of poverty appears to be breaking up.

Some implications of this dramatic breakthrough must be welcomed, but from an environmental perspective there are some obvious dangers. Measurement of our ecological footprint demonstrates that the biosphere cannot support even the present world's population living at a Western level of throughput. This is true even without the threat of climate change. Our direct impact on the hydrological cycle alone would force us to redesign our modern lifestyle in accordance with biospheric constraints. The scale of the problem means that no country can unilaterally resolve the issue and take care of its own citizens. This is a completely new situation in the history of the world, as far as we know.

The Kyoto Protocol to the UN Framework Convention on Climate Change has added further clusters to this new complexity of international relations, from OPEC (yesterday's fuel) to AOSIS (small island states – tomorrow's victims of climate change). China and the Group of 77 represent approximately what were the poorer countries, and they rightly insist that the rich countries must move first to reduce their carbon dioxide emissions before the rich can expect a commitment from the poor. Yet the Kyoto negotiations also brought a more explicit picture into focus: if the global community is to adopt a productive course of action then China and the USA – being, by far, the two largest emitters of carbon dioxide – must act together now. Both equity and pragmatism require this, because projections indicate that – on the present trajectory – by 2020 China will overtake the United States as the world's largest emitter of greenhouse gases (see Table 11.1).

We still face the same conundrum described in Chapter 1, that the poor countries cannot emulate the rich countries' economies – that is, follow the Western path – without ensuring that we, collectively, wreck the biosphere as a habitat for human beings (see Fig. 1.5). Even at the current stage of over-exploitation of sources and sinks we have set out on a completely unstable course, the result of which we cannot predict. Climate change

Table 11.1 The impact of China and the United States on the carbon cycle, current and cumulative

	China	USA
Population (millions) 1998	1243	270
Carbon emissions (million tonnes 1997)	885	1480
Carbon emissions per capita 1997 (tonnes)	0.705	5.303
Carbon emissions (billion tonnes) 1900–97	16.4	75.5
People per car	104	2

Source: Logan *et al.* 1999.

is the most dramatic and unpredictable aspect of human impact on the biosphere. But the fact is that it will only serve to reinforce large scale, biospheric changes that we have already set in train. Given these obvious dangers why would poor countries want to emulate the rich?

11.1.1 Modernism, colonialism and globalisation
The modernist movement dominated architecture and planning in the second and third quarters of the last century. Its origins can be traced to the Bauhaus Movement in which form followed function; starkness and simplicity reigned. The notion of the integrity of local style, tradition and value was swept aside by the triumph of the universal, based on modern materials and fuels. No matter where you were in the world you could build a skyscraper, served by elevators and chilled by air conditioning. Meanwhile the combustion engine, and later the jet engine, reduced the power of distance and gave the rich, at least, unheard of freedom to travel the world.

The political vehicle that permitted this cultural expansion was provided initially by European empires and later by American commercial power that began to operate on a global scale, unconfined by the old spheres of influence constructed by the Europeans. Local culture began to succumb on all fronts – political, economic and cultural. For a country to become 'developed' it was necessary to adopt and master the new Western technology – the new way of building, the new way of travelling, the new life in cities.

Colonialism provided the cloak for some of the transformation, but independence brought no change. If anything, the pace of the drive for modernisation, in the Western mould, picked up, especially in those countries that had something valuable to sell in the international marketplace such as oil. In order to participate in the modern world economy every country, however small and poor, needed an international airport, if not a national airline to go with it. The airport required a highway, a modern freeway with a minimum of four lanes, to sweep distinguished visitors into the city, to be deposited in a modern five-star hotel, complete with swimming pool and casino.

Globalisation has entrenched this development ideal. For successful members of the modern sector in Indonesia the expectations are similar to Illinois, or in Brazil, similar to Britain. A modern house, an automobile and international travel are all part of the package – in other words, a Western level of throughput. The rapid diffusion of the Western style of living has brought us to where we are today.

11.2 Health and climate change

11.2.1 Green and brown agendas
Despite the modern veneer, however, the condition of most large cities in the poor countries of the world is generally poor, with a high percentage of

people living in poverty, without a secure income, or access to clean water and sanitation (Stren and White 1989; Hardoy and Satterthwaite 1989; Hardoy, Mitlin and Satterthwaite 1992). On a daily basis they are still living with the original dangers implicit in urbanisation in the pre-industrial age. The major killers are still infectious diseases, such as malaria, measles and gastro-enteritis – all of which can be dramatically reduced with access to basic hygiene and adequate nutrition.

The environmental condition of the poor is sometimes summarised as the 'brown agenda', which, apart from urban problems, includes rural issues such as soil erosion (McGranahan and Satterthwaite 2000; White 2000; Zetter 2000). This problem set is sometimes contrasted with the supposed environmental preoccupations of the rich such as loss of biodiversity and atmospheric change, including climate change, which is referred to as the 'green agenda'. However – on a pragmatic level – these agendas cannot be separated because both need to be tackled simultaneously if we are to instil some balance into the rules governing the human occupancy of the planet. Indeed, in Chapter 9 I outlined a scheme that could use the concerns ascribed to the green agenda (principally climate change) to fund measures to address the issues of the brown agenda.

When we look at the large scale changes that are the focus of the green agenda we might think that the human species, as a whole, is already moving towards the edge of a precipice. However, the poor, still a majority of the world's people, are already living very near the edge. Climate change is going to push them closer to it, by drought, flood and hurricane.

11.2.2 Extreme weather events

Unpredictable water supply, more storms and floods, and higher temperatures will make life support systems more tenuous. The increase in temperature will bring more heat waves, more droughts and higher rates of evaporation from reservoirs and from the soil. All of these shifts are threats to food security. Lower river levels mean less power from hydroelectricity that could affect all kinds of service provision, such as pumped water distribution in cities.

The impact of the increased ferocity of cyclones and tropical storms has recently been witnessed (Table 11.2). In addition to the initial fatalities and economic loss, events of this magnitude have long term effects on human health and economic and social infrastructure. Drought – the most dangerous extreme weather event – is less dramatic than these storms, but the impacts are heavier. The 1999 drought in the United States killed 257 people and thousands of cattle. The 1998–9 drought in 11 states in India put 130 million people at risk. The same drought extended to Pakistan and Ethiopia. The combined impact of all extreme weather events, including droughts, is summarised in Table 11.3. Note the skewed impact on rich countries (represented by the United States) and on the world at large, which is

Table 11.2 The impact of major storms and floods in the South 1998–2000

Date	Location	Event	Deaths	People affected
June 1998	Gujarat, India	cyclone	10000*	n/a
November 1998	Central America	hurricane	10000	n/a
Summer 1999	Yangtse River, China	floods	n/a	2000000
Nov–Dec 1999	Vietnam	floods	700	1000000
October 1999	Orissa, India	cyclone	15000	
November 1999	Venezuela	heavy rains	50000	500000
January 2000	Madagascar & Mozambique	cyclone	n/a	n/a

Source: Swiss Re 1999.
* Munich Re 1999.

Table 11.3 Deaths and economic losses from storms and droughts, 1990–99, and 1999, United States and world total

United States				Worldwide			
1990–99		1999		1990–99		1999	
Deaths	Cost	Deaths	Cost	Deaths	Cost	Deaths	Cost
4000	200	450	14	330000	625	52000	68

Source: US Public Interest Research Group 2000.
Note: Cost represents economic losses in US$ billions.

mostly poor. For the 1990–9 period the United States suffered just over 1% of the deaths but more than 30% of the economic losses. The results would be further skewed in terms of insured losses that would lie between 20 and 30% of total losses in the United States, and would be almost negligible in the poorer countries.

11.2.3 Disease vectors and ultraviolet radiation
Although the health impacts of floods and droughts are heaviest in the poorer countries, the wider health implications of global warming are not good for the richer countries either, as food and water will be stressed there also. Furthermore it is expected that the temperature increase will generally favour the extension of disease vectors such as malaria, dengue fever, hantavirus and Lyme disease, in addition to bringing more heat waves. At the same time, the thinning of the stratospheric ozone layer is expected to

produce an increase in skin cancer and eye disease due to increased ultra-violet radiation.

11.2.4 Environmental refugees

Meanwhile, the persistence of poverty among the majority of the world's people will continue to generate an endless flow of environmental refugees as people flee from an intolerable present to some hope of a better future. The numbers that are moving now, at the very beginning of a new millennium, are a mere trickle compared with what could be produced by the climatic turmoil promised by the exponential growth in the emission of greenhouse gases. No defensive shield round a xenophobic Fortress Europe or Fortress America will be able to halt these movements. Almost all of these refugees will move to the major cities of the West, where – should they succeed in establishing themselves economically – they will use automobiles, central heating and air conditioning, and thereby dramatically increase their emissions of greenhouse gases.

11.3 Implications of the termination of the fossil fuel development model

These challenges to international co-operation and personal survival are occurring at a time when the fossil fuel development model must be phased out. Paradoxically this is a rare, even unique, moment of opportunity for the world's poorer countries. It is also a curious moment of tension, with a small percentage of the population quite well aware of the dangers of our current trajectory, while the rest are either completely unaware, or else they reject the information that is available to them. This situation was captured in President Clinton's speech quoted at the beginning of the chapter, which continued:

> Every member of Congress here will tell you that a huge portion of decision makers in our country and throughout the world – and most troubling, in some of the biggest developing nations – still believe that you cannot have economic growth unless you pour more greenhouse gases into the atmosphere. There is nothing so dangerous as for a people to be in the grip of a big idea that is no longer true. It was once true that you had to put more greenhouse gases into the atmosphere to grow the economy, to build a middle class, make a country rich. It is not true anymore. (US President Bill Clinton to the Democratic Leadership Council, 28 May 2000)

There is absolutely nothing to stop the poorer countries following the industrial model of the rich. They can build their own power plants and automobiles. They can train their own technicians and raise their own capital. Many already do so; foreign companies are investing in poor countries to boost profits, not to 'transfer' some mystical technical knowledge. Under this scenario – labelled 'business as usual' by the Intergovernmental Panel on Climate Change – fossil fuel emissions will continue to rise exponentially, with consequences that are somewhat unpredictable, but certainly not very conducive to human comfort.

This view – that climate change is a serious threat that requires action right now – has become widely accepted only since the Kyoto Protocol was signed in November 1998. Right until the last minute of the negotiations it was unclear that the diplomats who drafted the protocol would agree that the likely consequences of business as usual would be seriously adverse for the human population. They still had to convince politicians, business leaders and citizens at home. (The signatories to the Convention – known as Annex 1 countries – are listed in Appendix 3 in this book.) This process is moving slowly. What is significant is that the most vocal and wealthy opposition to any climate agreement – the Global Climate Coalition – subsequently began to fall apart, as one member after another has resigned from the group. The Coalition was formed by large manufacturing (chemicals, automobiles), fossil fuel and mining companies and industry associations specifically to deny that greenhouse gas emissions might be a serious problem (Leggett 1999).

The rich countries of the world must now determine just how they will make the first round of carbon dioxide emission reductions to which they committed themselves at Kyoto. The Bush administration has apparently withdrawn the United States from the Kyoto process, but the physical problem of climate change remains. The Kyoto agreement described various mechanisms such as the Clean Development Mechanism, Joint Implementation, and carbon trading that would allow the richer countries to gain emission credits for reducing emissions in poor countries, or in other countries that were signatories to the Protocol. It was not specified *how much* of their cut could be made in these ways. As President Clinton indicated, to earn any international credibility the rich countries will have to make some real reductions at home. In fact, progressive businesses have already begun to do so and momentum is growing every day (Romm 1999; Sandor 2000).

When rich countries have developed a strategy for reducing carbon dioxide emissions, commitments will be needed from the poorer countries, especially the major industrial powers and population centres like China, India, Indonesia, Mexico and Brazil. Otherwise the efforts of the richer countries will be wasted if the poorer countries continue with business as usual. There is only one peaceful incentive that the rich countries can offer to the poor to restructure their economies and societies in order to adapt to a low carbon lifestyle. That is one in which emission rights are distributed on a per capita basis, with national populations fixed to some base year such as 1990 – the base year used for pegging carbon emission reductions. Some of these rights might then be traded from poor countries to rich, or other frameworks for resource transfer might be established – as suggested in Chapter 9.

11.4 Potential for cities to lead the way

More than half the world's people now live in urban areas and that number is still growing. More importantly, cities have both the power and the motivation to move swiftly to reduce their greenhouse gas emissions. They have the power because they manage transportation and land use issues within their boundaries, and there are a host of other policies they could pursue, as described in this book. They have the motivation because they could simultaneously tackle a great many related problems such as air quality, traffic congestion, water delivery and power delivery. Anyone working at the local level understands that these issues are linked on the street, even if they are not linked in the various ministries of our national governments. Furthermore, the people who run urban governments – both civil servants and elected politicians – live much nearer to the daily lives of the people they serve than do politicians who work at the national level where the interests of the entire country must be weighed together.

Also – if further reason were needed – cities are extremely vulnerable to climate change and other associated biospheric challenges, such as water scarcity and food security. Modern cities produce hardly any of their own water, food or power. They may generate electricity, but the fuel – like all the other supplies – is brought in. If supplies are interrupted they have virtually no capacity for self-reliance.

Despite all these good reasons why cities should be in the forefront of the climate change negotiations, they are invisible in the debate, because climate change has been approached as an international problem – which it certainly is. But many – if not most – of the solutions lie in the cities.

The virtual exclusion of urban governments from the climate change negotiations is not the only evidence of a general lack of understanding of the great potential for transforming our urban settlements into ecological cities. The scientific forum for developing climate change policy is the Intergovernmental Panel on Climate Change. In the work of the Panel, cities are scarcely mentioned; certainly their potential contribution to solutions – as governments, and as communities – is ignored. Even the documents the Panel has prepared on technology transfer (from rich countries to poor, naturally) ignore cities as ecosystems. 'Technology' is presented in the traditional sectors – water, food, transportation, power and so on.

Despite this policy of exclusion from both the negotiations and the development of the science, cities have an enormous potential to lead the way on climate change, as I hope has been demonstrated in this book. The very concentration of people and activities that epitomises the city presents opportunities for more efficient use of resources and sinks. The technology already exists to implement significant improvements in a cost-effective way. However, it will be difficult to generate the necessary grass-roots support. Financial incentives are needed that will make high carbon emitters pay for the cost of developing a low carbon lifestyle, while rewarding those who continue to emit very little.

People may think that each city is only a small part of the environmental problem, yet each success is a demonstration that something works – and this might be replicated in other cities. Like an antibody in a diseased organ, the ecological city can cleanse the rest of the system. This potential for replication links the cities of rich countries with cities in poor countries because they are relying on the same technology, and it is that technology and the lifestyle associated with it that must change. The way in which more environmentally benign urban systems come into being is something that will be worked out pragmatically on the street, not in the international negotiating fora.

There is resistance from national governments to accept urban governments as full partners in this process. That is one reason why progress has been slow. However, as cities develop their potential it will be a foolish national government that continues to muddle on without accepting their active participation in policy formulation and implementation.

11.5 International urban co-operation

The principal association to promote exchange of information among urban governments on a global scale is the International Union of Local Authorities (IULA) headquartered in The Hague. The IULA holds a biannual meeting and provides administrative and technical training for the staff of local authorities throughout the world, including developing countries. It has no explicitly environmental mandate, although, implicitly, urban services are all about environmental quality.

It was not until the foundation of the International Council for Local Environmental Initiatives (ICLEI) in Toronto in 1990 that urban governments had the opportunity to focus on urban environmental issues within the context of climate change, and hence identify strategies to protect themselves and the world at large from its expected impacts. Other international environmental organisations have a strong influence in urban issues, including the International Institute for Sustainable Development in Winnipeg, and the International Institute for Environment and Development in London.

ICLEI works in co-operation with IULA and the United Nations Environment Programme, and was responsible for the Nagoya Conference which was held by urban governments, in parallel to the Kyoto Conference. The outcome of that conference is known as the Nagoya Declaration, which is reproduced in full as Appendix 2 in this book. There were several significant differences between the events in Kyoto and the events in Nagoya, despite the fact that the two meetings were debating the same issues at the same time in the same country. Kyoto was a bitter struggle between adversaries to reach a rather limited compromise. Nagoya was a consultation among partners about confronting the problems that they faced together. It is revealing that in their opening paragraph the Nagoya parties noted the IPCC opinion that 'stabilization of the global climate may require reductions in greenhouse gas emissions of more than 50%' (see Appendix 2, Preamble). They also recognised that reductions in greenhouse gas emissions would produce multiple benefits such as:

- lower costs of municipal operations
- local job creation
- better air quality and improvements in public health
- reductions in traffic congestion
- better urban liveability (Nagoya Declaration, paragraph 1.6).

An important difference between the meetings was that the Nagoya group included urban governments from poor countries as well as rich. The subsequent paragraph explains the reason for their participation:

> Inspired by the multiple benefits that local governments in developed nations are enjoying as they successfully implement greenhouse gas emissions reduction strategies, local governments in developing nations have also begun to initiate actions to reduce greenhouse gas emissions, even though their national governments are not obligated under the Framework Convention on Climate Change to reduce such emissions. Provided adequate financial resources and appropriate technologies are available, developing country cities are desirous of pursuing sustainable development strategies that reduce greenhouse gas emissions. Local governments in developing nations believe that when strategies to reduce greenhouse gas emissions address social problems and improve the standard of living locally, they are indeed worth pursuing. (Nagoya Declaration, paragraph 1.7)

In one compact statement, the delegates explained exactly how poorer countries could be involved in a global strategy to reduce greenhouse gases, while attending to related environmental issues as well. There is no distinction between green and brown agendas here because it is recognised that there is only one agenda after all. The Nagoya Declaration closed with the following Local Government Pledge:

> WE, the representatives of local governments worldwide, are convinced that local governments are an effective vehicle to help achieve national greenhouse gas reduction goals and that close partnerships with our national governments will enhance our collective efforts to reduce greenhouse gas emissions;
> THEREFORE, WE pledge to make maximum efforts to reduce greenhouse gas emissions at the local level, to embrace the goals of global sustainability and Local Agenda 21, and to extend our full support to our national governments and to ICLEI's Cities for Climate Protection campaign, which will seek to:

3.1 Promote the significant reduction of greenhouse gas emissions from municipal and other public operations, including buildings, facilities, landfills, waste treatment, and water pumping stations through the use of renewables, energy efficiency, co-generation, district energy and recycling.
3.2 Promote the significant reduction of greenhouse gas emissions from community-wide activities, including transportation, housing, and commerce locally.
3.3 Promote the significant expansion of the supply and consumption of renewable energy worldwide at the municipal level.
3.4 Promote local educational initiatives and organisations to enhance public understanding of climate change, thereby improving acceptance of national government measures to reduce greenhouse gas emissions.
3.5 Recruit by the year 2000 local governments to the CCP campaign that together with existing CCP participants account for 10% of the world's carbon dioxide emissions.
3.6 Significantly expand the CCP campaign in developing countries.
(Nagoya Declaration, paragraph 3)

11.6. Emissions trading – the lure and the danger

An argument was made in Chapter 9 that any scheme to begin trading in emission permits should include cities, as they have both the means and the motive to make significant reductions right now. Cities are collectives of households, infrastructure and companies, each of which might understand the benefits of reducing energy consumption and switching to lower carbon sources of energy.

At this point we need to examine some of the international implications of emissions trading and the other 'flexible mechanisms' included in the Kyoto Protocol, because the trading proposal was an outgrowth of the earlier approaches. Soon after the Rio Conference was concluded and the Framework Convention on Climate Change drafted, the observation was made to the effect that it made little economic sense for efficient economies (i.e. OECD economies) to spend money reducing their greenhouse gas emissions so long as more reductions could be achieved for the same investment in less efficient economies (i.e. the poor). The concept of Joint Implementation (JI) was born. Under JI bilateral partnerships were formed between rich and poor countries (e.g. Canada and India, Japan and Indonesia) to implement energy-efficient operations in the poorer partner's country. The concept soured when it was suggested that such rich country efforts should be taken as a credit towards any emission reductions the rich committed to make. The obvious danger was that the rich would not reduce their own emissions, while JI would make barely a dent on the growth in emissions in poor countries. The Clean Development Mechanism was a slightly more subtle approach, which paid the poorer countries directly for administration of the projects and put the management under multilateral control.

Emissions trading was founded on the same logic that it made more 'economic sense' to invest in reductions in the most energy-inefficient producers, only this time the trading was restricted to Annex 1 countries, the richer, industrial signatories to the Kyoto Protocol, excluding the poor of the developing countries. The proposal would allow an 'efficient' economy, like the United States, Germany or Japan, to purchase 'reduction credits' from less efficient producers. However, permits and targets were based on 1990 figures, and Annex 1 included the countries from the former Soviet Union and ex-communist countries of Central Europe (see Appendix 3). All of the latter experienced economic implosion from 1990, with attendant, inadvertent reductions in greenhouse gas emissions. Under the logic of the emissions trading proposal the richest countries could buy these 'hot air' reductions from their poorer industrial neighbours and use them as part of their 'reduction commitment'. Like JI and CDM the scheme had the potential for playing plenty of games while doing nothing for reducing greenhouse gas emissions.

It is likely that the flexibility mechanisms will play some role in the implementation of the commitments made under the Kyoto Protocol. They have the potential to encourage communications between the national players and they may genuinely help to reduce emissions. Equally clearly, they should not be allowed to make up 100% of any country's Kyoto commitment because of the dangers of the richest countries making few real moves to reduce emissions. As President Clinton pointed out in the opening quotation to this chapter, the richest countries must 'demonstrate that we have evidence that a different way will work'. The evidence must be based on actions that have been taken in the transportation, production and urban systems of the richest countries themselves.

11.7 Conclusion

11.7.1 Elements of a coherent response

International relations have turned out to be a difficult starting point for establishing an agreement on redefining guidelines for human interaction with the biosphere. Given the vast discrepancy between the rich and the poor, and given that the recent accumulation of wealth is highly correlated with the use of fossil fuels, the difficulties were predictable. The proposed retreat from fossil fuel use could not have come at a worse time for the poorer countries that have only just got to the point of benefiting from modernising their economies. There is a lot of coal in China and India, and a lot of oil in Indonesia. Are they expected to leave it in the ground?

There is some chance that another route to agreement, based on mutual

benefit, can be found by working at the local level, where urban governments can work as equal partners, dealing with similar problems. Water is
water; a house is a house. There is a growing body of evidence that suggests
than an alliance of concerned citizens, progressive companies and cities with
a vision can go forward to a future that makes ecological sense. A review
of how this might be done is the subject of the final chapter.

11.8 Websites

1. Aalborg Charter of European Cities and Towns Towards Sustainability,
   at: *http://www.iclei.org/europe/echarter.thm*

2. Beijing Environmental Protection Bureau:
   http://www.bjepb.gov.cn/English_homepage/index.htm

3. Cities Environmental Reports on the Internet (CEROI):
   http://www.grida.no/city/projdes.htm

4. European Commission for Local Sustainability:
   http://cities21.com/coldfus/citylist.dbm

5. International Council for Local Environmental Initiatives:
   http://www.iclei.org/

6. International Institute for Sustainable Development: *http://www.iisd.ca/*

11.9 Further reading

Grubb, M with Vrolijk, C and Brack, D (1999). *The Kyoto Protocol: A Guide and
   Assessment*. London, Royal Institute of International Affairs.
Hardoy, J, Mitlin, D and Satterthwaite, D (1992). *Environmental Problems in Third
   World Cities*. London, Earthscan.
Meyer, A and Cooper, T (2000). 'Why convergence and contraction are the key',
   Environmental Finance, May: 19–21.
Stren, R E, White, R R and Whitney, J B, eds (1992). *Sustainable Cities: Urbaniza-
   tion and the Environment in International Perspective*. Boulder, Westview Press.
White, R R (1993). *North, South and the Environmental Crisis*. Toronto, University
   of Toronto Press.
Wolfensohn, J D (1996). 'Crucibles of development', *Human Settlements – Our
   Planet*, 8.1. Nairobi, United Nations Environment Programme.
World Bank (2000). *Entering the 21st Century: World Development Report
   1999/2000*. New York, Oxford University Press.

12

Do we have the means to build the ecological city?

We, cities and towns, recognise that a town or city cannot permit itself to export problems into the larger environment or to the future. Therefore, any problems or imbalances within the city are either brought towards balance at their own level or absorbed by some larger entity at the regional or national level. (Charter of European Cities and Towns Towards Sustainability – The Aalborg Charter 1994: para 1.5)

The fate of urban areas could go either way: they could become human and environmental disaster areas, or they could become centres of global creativity, prosperity and growth of the human spirit. In many ways, how we approach urban development is central to the broader challenge of development, because cities concentrate all problems and opportunities. Working in partnership, I am convinced that we can ensure that the cities of the 21st century become arenas of opportunity and hope for mankind. (Wolfensohn 1996: 1)

12.1 An ecological city: what does it really mean?

This chapter reviews the potential for a modern, Western city to become an ecological city. For better or for worse, the Western city is the model for cities in developing countries and emerging economies. It is therefore important for all of us to understand the ecological flaws of the Western city and to devise the necessary countermeasures in order to restore balance to our biospheric niche.

Do we have the means to build an ecological city? Yes we do. At least we have the technical means. What we generally lack is the political imagination, political will and personal motivation. We can blame institutions and institutional leaders, but it is at the individual level that we have failed to understand our problem and take appropriate action.

The definition proposed at the beginning of this book was that an ecological city was one that 'provides an acceptable standard of living for its human occupants without depleting the ecosystems and bio-geochemical cycles on which it depends' (page 1). This was something that we did naturally – until we discovered coal, and then let the fossil fuel genie out of the bottle. Since then we have vastly overshot our limits. Our current task can be stated quite simply: we have to bring ourselves back into balance with what the planet can support, as recognised by the signatories to the Aalborg Charter, quoted above. (The full text of the Aalborg Charter is provided in Appendix 1.) We are the head of the food chain, so if we destroy the components on which we depend, then ultimately we will become victims of our own stupidity.

This simple definition of an ecological city, and hence our proposed global objective, should not obscure the fact that there are many important questions that this book has passed over. For example, from time to time I have noted that we need to decide at what spatial scale an environmental problem can be addressed. In the opening quotation from the Aalborg Charter it states that towns and cities should not 'export problems into the larger environment'. Similarly, elsewhere we have noted that cities should take care of 'their own' wastes. Yet moving from aspiration to action on these problems requires an understanding of the emerging field of political ecology which raises difficult problems of responsibility in the spatial and ecological context (Atkinson 1991). These depths have not been probed in this book.

On the contrary, this is a very simple, empirical book. As guidelines, it proposes that we should aim for a much reduced input of resources and the elimination of wastes – overall a much reduced throughput, a much smaller footprint. What this means, roughly, at the present population level, is per capita goals of 500 kg of carbon output per year, 200 litres of water used per person per day, and zero solid waste. By setting these local goals we can achieve the global goals required to stabilise our niche. 'Thinking globally, acting locally', as adopted by the Stockholm Conference on Environment and Development in 1972, was exactly the right motto.

In order to achieve this goal we should sketch out a programme for the short, medium and longer term, meaning approximately two, ten and twenty years respectively (see sections 12.4, 12.5 and 12.6). Before that it will be instructive to review some lessons learned from the impact of recent extreme weather events on cities (see sections 12.2 and 12.3). These studies enable us to see the city in a holistic situation where every aspect of daily life comes under pressure at the same time. These lessons relate only to the adaptive role that cities can undertake to respond to more extreme weather events associated with climate change. They do not cover the mitigative role that cities can play by reducing greenhouse gas emissions, which has been one of the main thrusts of this book.

12.2 Urban lessons learned from recent extreme weather events

12.2.1 General lessons

As always, prevention offers the biggest return on investment. Much of the damage that households and businesses suffer from extreme weather events could be prevented if regulations are enforced, especially with respect to land use zoning and building regulations. However, what is needed is more than regulation of specific actions. As David Etkin remarked: 'Current thought emphasizes a more holistic approach that incorporates social, political, environmental, economic and cultural considerations as well as technological ones. It is worthwhile noting that this change has taken a long time to become accepted' (1999: 72).

Accurate weather prediction is the next major factor. Some weather phenomena like heat waves and hurricanes take days to build up. Others like tornadoes and hailstorms give only moments' warning. From the urban management perspective the key is to take predictions seriously and to mobilise the necessary response. For the authorities this response must be integrated, both by sector and spatially. The eventual impact of the weather event can be reduced by strong community spirit, capable of a certain amount of self-reliance.

Under severe weather stress, assets may become liabilities. For example, subways may be flooded, trees may come down, and automobiles become immobilised in the street, blocking the passage of emergency vehicles and fleeing residents. The most vulnerable technological aspect of a modern city is its dependence on an integrated electricity network. Problems at a single point in the network can result in widespread outages. If extreme weather events become more commonplace then a modern house (and other buildings in a city) will need some source of back-up power, perhaps using the kind of uninterrupted power supply (UPS) technology that is currently found only in certain commercial operations, such as banking, where loss of power can be very costly (see section 3.5).

12.2.2 Heat wave: Chicago, July 1995

The build-up of the heat wave was predicted days in advance but few preparations were made in Chicago. Emergency air conditioned tents were set up but many elderly people in the poorer neighbourhoods lived alone and were afraid to leave their homes. Not only did they live alone but they obviously had no real contact with their neighbours. Perhaps the most appalling statistic to come out of this event was the number of unclaimed bodies that eventually were buried in a common grave. These people had been alone in their life and were abandoned in their death. Such vulnerable people provided the pool for most of the fatalities from heat-stroke. The widespread failure of the power system shut off air conditioners and fans at the height of the event.

Since this tragedy two major efforts have been undertaken to make Chicago a greener place and therefore perhaps less susceptible to extreme heat. Brownfield sites are being redeveloped, some as gardens and parks (see section 8.5). There is also a campaign to develop an urban forest canopy. In other large North American cities people asked: What would happen here under similar weather conditions? How many deaths would we sustain?

12.2.3 Ice storm: Montreal, January 1998

Emergency preparedness in Montreal had been focused on small scale events, such as an aircraft missing the runway of the nearby airport and striking the city, or a terrorist attack. Meanwhile the great vulnerability of the city was its dependence on just three overhead power lines carrying the power supply for the island of Montreal, containing the business hub of the entire urban system, and serving over 2 million residents and daily commuters. Although the onset of the ice storm was predicted, its unusual duration was not. The build-up of ice on the power lines surpassed all experience and the pylons began to fall down, cutting off power to many parts of the city, including power for pumping water. At one point there was only four hours of water supply left for the city. A fire outbreak at that time – easily sparked by a collapsing electric cable – could not have been contained. The paradox of the ice storm is that the greatest risk was fire.

Although there are 15 road bridges off the island, most of them were blocked by ice, stalled vehicles and fallen trees. Most of the residents were essentially trapped. Even if they could have been taken off the island there was nowhere available to house that number in the middle of the Canadian winter. In their homes people used paraffin heaters with inadequate ventilation; some died from asphyxiation as well as small household fires.

The Canadian insurance industry made record payouts exceeding 1 billion dollars, the largest ever made for a single event in Canada. Payments were prompt and helped to stimulate the economy as people repaired the damage. However, for an event of this magnitude the industry might have to consider being more flexible. For example, payments were not made to clear ice from the roofs of buildings, as this was the owner's responsibility. Some owners let the ice accumulate and buildings collapsed; others attempted the clearance themselves and suffered injuries and even some deaths. Similarly insurers would not pay the transport costs of people leaving town for safety, yet it was not possible to move to local hotels (which could be covered by insurance) because they were either full or were inoperable because of lack of power or water.

12.2.4 River basin floods: Grand Forks, North Dakota, and Winnipeg, Manitoba, April 1997

Flooding is the greatest single danger for cities under climate change. The enhanced hydrological cycle means that all previous estimates of the flood hazard must be updated. Adherence to zoning regulations is crucial to keep new encroachments out of the flood plain. Existing buildings in the flood plain should be cleared as soon as practicable. The 1990s have witnessed major flooding around the world.

As for the ice storm, there is a paradoxical risk of fire, as was seen in Grand Forks, North Dakota, which lost four blocks to a fire in the central business district (Partner Re 1997). The centre became an island in the floodwaters produced by a rapid snow-melt, making access for fire engines almost impossible. Grand Forks, a town of 70 000, was evacuated before it was submerged. Subsequently, the Red River, on which Grand Forks is situated, flooded down towards the still frozen waters of Lake Winnipeg, surrounding the City of Winnipeg on its way. Through several tense weeks the dyke surrounding the city protected it from the flood which covered the surrounding farmlands and small towns.

12.2.5 Hurricanes, tornadoes, hailstorms

Just as adherence to zoning regulations is the key to reducing losses from flooding, so is respect for the building code the key for reducing the impacts of windstorms, and heavy rainfall. Although post-storm photos concentrate on flattened buildings, most of the damage to buildings is caused by a rupture of the building envelope (usually the roof) followed by incoming rainfall. If the building is built according to code the rupture is very unlikely to happen. The insurance industry in storm-prone North America is now taking a pro-active line by offering a discount on insurance premiums where local authorities really enforce the code.

In addition to the impact from the wind and rain of the hurricane itself there are additional costs inflicted by tornadoes spawned by the hurricane, by storm surges along the coast, and by inland flooding. Inland flooding is created by a combination of prolonged heavy rainfall and the backing up of rivers by the storm surge. In the United States, where more than 60% of the total population lives in coastal and island states exposed to the hurricane threat, more than half the deaths over the last thirty years have been due to inland flooding (*http://www.nws.noaa.gov/oh/hurricane/inland_flooding.html*).

The very size of hurricanes, the slowness of their progress, and the unpredictability of their point of landfall means that hurricanes generate the largest number of evacuees. Hurricane Georges (September 1998) caused the evacuation of downtown New Orleans, although it eventually made landfall 60 km to the east. Hurricane Floyd (September 1999) passed closely along the coast of Florida, Georgia, South and North Carolina before making landfall at Cape Fear (NC), by which time an estimated 3.2 million people were on the road. The simultaneous evacuation of so many people by automobile predictably produced gridlock and the people would have been defenceless if the storm had swept inland. Ironically, even as the vehicles sat in the traffic jam, people kept the motors on to run their air conditioners – all the time pumping more carbon dioxide into the atmosphere, until their petrol was used up.

The authorities congratulate themselves on the small loss of life and injury, given the scale of such operations, but these are very unstable situations socially and politically, and it is only a matter of time before one goes badly wrong even in a technologically developed culture. Some people refuse to leave the path of the storm. As in the case of the Chicago heat wave, it is usually the poor and the elderly who are most reluctant to leave. Those who are dependent on the authorities for evacuation may later protest violently, as happened during hurricane Georges. On that occasion 14000 people were taken to the nearby sports stadium for shelter and kept there for 15 hours by armed National Guardsmen.

12.2.6 Conclusion: extreme weather and the ecological city

There are several specifically ecological reasons why an ecological city would be in a better position to withstand extreme weather events. There are also some general principles of good management and community solidarity that should help and might be expected to be an intrinsic part of an ecological city.

From an ecological perspective an extensive tree canopy and vertical gardens will reduce air temperatures and hence reduce heat stress for humans. The encouragement of natural, porous surfaces and green infrastructure should absorb heavy rainfall as well as reduce air temperature. The availability of multiple power sources – such as solar PV in houses and other buildings – will reduce vulnerability to loss of power. Power also supports essential public services (especially water supply, telephones, radio and light) and makes it possible for individuals to stay in their homes and for businesses to keep running. A modal split tilted towards walking, cycling and public transport is more resilient because these modes require less road space to move a given quantity of people. In an emergency, automobiles are more likely to clog the roadways and are more vulnerable than buses to flowing water and damaged roadbeds. Furthermore, public transport is run by professionals, with access to radios and trained in emergency procedures. In comparison, from a mass transportation perspective, automobile drivers are amateurs with no capacity for co-operative action.

Good management and community solidarity should support key preventive measures such as adherence to zoning regulations and the building code, and the development of effective emergency response procedures. The downside of an automobile-based society is that you are less likely to know your neighbours than you are in a society where cycling, walking and public transit are the norm.

12.3 Case study: American cities at risk

As part of its programme on Cities for Climate Protection, the International Council for Local Environmental Initiatives assessed the vulnerability of five American cities to global warming with respect to health, municipal infrastructure and the local economy. The results are summarised in Table 12.1. What the table illustrates is the local nature of adaptation that will be required to meet the generic threat of climate change. It does not include the more widespread health risks due to changes in disease vectors, for example. Also the list is limited to direct impacts on the economy, particularly the physical impacts on tourism and the agricultural sector, including

Table 12.1 Potential impacts of global warming on five American cities

	Health	Municipal infrastructure	Local economy
Boston	ozone,* heat stress	water shortage, storm drains and sewage back-up, flood risk for airport, subway and sewage treatment plant	fishery decline maple tree decline
Chicago	ozone, heat stress	adapt water supply and drainage to lower lake levels	entrepôt for regional agricultural economy subject to climate stress
Denver	ozone, heat stress	water shortage, impact of winter storms on airport functioning	regional agriculture, winter sports
New Orleans	maybe little impact	major threat of flooding from hurricanes and sea level rise, saltwater intrusion of the water table	fisheries, especially shrimps vulnerable to warming
Seattle	little or no impact	risk of flooding and water shortages, risk from sea level rise	agriculture, forestry and fishery all vulnerable

Source: ICLEI 2000, summarised from pp. 26–34.

* 'ozone' refers to excessive concentrations of ground-level ozone with attendant risk for pulmonary function, especially asthma.

fisheries and forestry. Any negative impacts on these sectors would have a ripple effect through related activities in finance, transportation, wholesaling and retailing.

These cities are not among the most vulnerable on the planet. They are not sinking like Venice and London; they are not the home to millions of already marginal people living in shantytowns in the path of tropical storms. Yet all face an increase in vulnerability. Air quality is likely to deteriorate in Boston, Chicago and Denver. All five of them will suffer hydrological disturbances that may bring water shortages. Downtown New Orleans is below sea level, while Boston's airport runs out along a promontory barely two metres above the waves. Climate change will affect all their agricultural operations and forestry, as well as fisheries for the coastal cities. All of them are highly automobile-dependent, which is one of the primary causes of the problem.

12.4 Achieving what is already within our reach

In the short term we can make a list of changes that can be made to urban systems using existing technology that is already operational somewhere in the world, such as the checklist shown in Table 12.2. The Local Government Management Board (UK Department of the Environment, Transport and the Regions) produced this list, and it looks fairly innocuous. Similar lists could be drawn up in any Western country. They may look harmless and – as this subtitle suggests – 'within our reach'. The first goal is to 'use energy, water and other natural resources efficiently and with care', although the guideline 'efficiently and with care' is wide open to interpretation. If we go back to the diagram of energy flows in Toronto (Fig. 3.2) we might be prepared to agree that 50% efficiency was not high for a large urban system. As we saw in Chapter 6, it was not so difficult to reduce greenhouse gas emissions by 6% in the Toronto–Niagara Region. But is this really 'efficient'?

The next goal – limit pollution to levels which do not damage natural systems – is much more specific, if we take 'pollution' to include greenhouse gases and if we take 'natural systems' to include the composition of the atmosphere. Under this regime we would cease emitting excess greenhouse gases at once. Later, under the subtitle 'promote economic success' the checklist suggests local authorities 'encourage access' 'in ways which make less use of the car'. This goal is considerably milder than eliminating excess greenhouse gases but it still implies a reversal of present trends, both in the UK and in the United States and Canada, in all of which countries the use of the car is increasing.

Thus, even this innocuous list poses challenges that we are not yet able to meet on a national scale, despite local successes. The reasons for this failure are not hard to find. As noted in section 9.1, our incentive structures

Table 12.2 Characteristics of a sustainable society: checklist for local authorities

A sustainable society seeks to:

protect and enhance the environment
- use energy, water and other natural resources efficiently and with care
- limit pollution to levels that do not damage natural systems
- minimise waste, then reuse or recover it through recycling, composting or energy recovery, and finally sustainably dispose of what is left
- value and protect the diversity of nature

meet social needs
- create or enhance places, spaces and buildings that work well, wear well and look well
- make settlements 'human' in scale and form
- value and protect diversity and local distinctiveness and strengthen local community and cultural identity
- protect human health and amenity through safe, clean, pleasant environments
- emphasise health service prevention action as well as care
- ensure access to good food, water, housing and fuel at reasonable cost
- meet local needs locally wherever possible
- maximise everyone's access to the skills and knowledge needed to play a full part in society
- empower all sections of the community to participate in decision-making and consider the social and community impacts of decisions

promote economic success
- create a vibrant local economy that gives access to satisfying and rewarding work without damaging the local, national or global environment
- value unpaid work
- encourage necessary access to facilities, services, goods and other people in ways that make less use of the car and minimise impacts on the environment
- make opportunities for culture, leisure and recreation readily available to all

Source: adapted from Local Government Management Board (UK) 1994.

are still skewed towards subsidising fossil fuels and nuclear power. We have the means to implement the goals on this list right now, but few people see the need to do so yet.

12.4.1 Cities for Climate Protection Campaign

ICLEI's Cities for Climate Protection Campaign is built on a network of nearly 400 local communities around the world, from rural districts to the largest cities such as Calcutta, Chicago, Berlin and Rio de Janeiro. In income it ranges from Aspen, Colorado, to Zomba, Malawi. The network serves as a clearing house for best practice case studies that deal with a range of environmental problems, including mitigative measures for climate change. Table 12.3 summarises key activities in two cities: Bologna, Italy, and Portland, in the north-west United States.

Table 12.3 Strategies for the reduction of carbon dioxide in Bologna, Italy, and Portland, Oregon

---

A. CITY OF BOLOGNA, ITALY

Land use and solid waste

1. Materials recycling:
   - construct waste incinerator to supply district heating
   - burn biogas from landfill
   - heat recovery from sewage treatment plant
2. Reforestation:
   - increase biomass on public green space by planting trees and bushes on grassed areas
   - expand public green space

Air quality and energy

3. Transportation:
   - increase efficiency of public transit system
   - rebuild light rail system in the city, in conjunction with improvement of regional rail system
   - reduce circulation of private cars in the urban core
   - bus passes for municipal workers
4. Urban planning:
   - enhance public transport
   - extend district heating
   - extend building code for energy conservation
5. Energy saving:
   - reduce consumption in hospitals, universities, residences, municipal offices
   - extend distribution network for natural gas
   - cogeneration for new residential areas, new hospital complex
   - energy retrofit for municipal buildings
6. Renewable energy:
   - develop hydroelectric potential from canal system
   - install passive solar heating for water in sports complex

Education

7. Information campaign through new Office for Energy and Environment to reach:
   - commercial enterprises
   - citizens

B. CITY OF PORTLAND, OREGON

Land use and solid waste

1. Recycling:
   - increase recycling rate from 26% to 60%
   - purchase paper with minimum 25% post-consumer waste content
2. Reforestation:
   - extend urban tree planting and improve urban tree maintenance
   - encourage state-wide reforestation

Air quality and energy

3. Transportation actions to reduce vehicle miles travelled:
   - expand transit system, including light rail
   - promote more compact development
   - make streets bicycle and pedestrian friendly
   - adopt fees to reflect full cost of driving
   - purchase fuel efficient vehicles for the city
   - better management of road accidents to reduce delays
4. Energy efficiency:
   - reduce residential, commercial and industrial use by 25%, 20% and 15% respectively
   - use methane from landfills and waste water treatment plants
   - promote renewable energy, district heating and use of waste heat

---

Source: ICLEI 2000. Details available from *http://www.iclei.org/aplans/bolognap.htm* and *http://www.iclei. org/aplans/portlap.htm*.

Despite the differences in history, culture, climate and site, the proposed strategies have much in common and include many of the approaches described in this book. All efforts are targeted at reducing throughput – of energy and materials. (As the general goal, for this exercise, is carbon dioxide emission reduction, there is no explicit reference to water management.)

12.5 Changing gear, or moving the game up to another level

Cities in the richer countries can implement the measures described above, even in the existing political and economic context. Everything we need in terms of technology is already available, lying to hand. All we have to do, in the richer countries, is to put it together. However, the opportunities for Third World cities taking similar steps depend heavily on reaching an international understanding on climate change, as set out in Chapters 9 and 11.

All cities could achieve far more if we change the nature of the game we are playing and set ourselves higher standards. The current incentive structure is about making money and paying taxes. Pollution – the production of harmful residuals – is punished only if there is a gross threat to human health. There are hardly any incentives to care for the earth, create meaningful work for members of society, or to reflect on where our society is heading. The people who pursue these goals are a minority who work against the tide of social change. The wealthier we become as a society, the shorter are our time horizons. Yet the rate of technological change is ever increasing. Today's invention becomes tomorrow's ubiquitous product.

There is a better game that we could play. We could actually change the incentive structure so that society as a whole – or at least the majority, rather than the small minority – would work towards those goals that we say are desirable, but which seem so hard to achieve. We could replace incentives to pollute with incentives to restore and conserve. We could link environmental needs to social equity. We could establish per capita rights to resources and sinks and allow people to trade them at their real biospheric value. This is the new kind of agreement that would allow all cities – rich and poor – to undertake significant steps to put their systems back into balance with the natural world. Such an agreement would be based on a firm understanding of the targets, or endpoints, that we should aim for in order to stabilise the major biospheric parameters on which we depend. Any rights to resources and sinks should be based on that understanding. Such rights would need to be regulated and supervised by a World Carbon Council, a body with a more significant role that its purely economic precursor, the World Trading Organisation.

As Ivan Head observed, this step towards recognition of individual rights on the global scale really places us 'On a Hinge of History' (1991). It is not at all clear whether we are capable of developing a shared understanding of our predicament. After the failure of the Convention of the Parties to the Framework Convention on Climate Change at The Hague (November 2000) it seemed unlikely that we could transcend a narrow, regional vision based on perceived self-interest. Mentally we are not well equipped for such a transformation despite increasing evidence of the need. Edmund O'Sullivan identifies the need for 'transformative learning' because:

> In spite of the fact that there are regional differences across this planet that would indicate there are differences in responsibility for our present planetary concerns, we nevertheless know that there are vast problems facing this planet as a whole. (O'Sullivan 1999: 18)

At this stage we can only speculate on which path humanity will take.

12.6 Getting ahead of the problem

We will get ahead of the problem only when people get used to living within their biospheric means. Only then will it be possible to develop a global dialogue directed towards stabilising the impact of human society at a level that is sustainable. To do that we need to bring the major bio-geochemical cycles back into balance, especially the carbon cycle and the hydrological cycle, which are the ones that pose the most imminent danger. This is what is implied by the term 'sustainable development'.

By 2020 perhaps 70% of the world's population will be living in cities, linked by a communication network that allows them to compare results from a variety of technologies. As stated at the beginning of this book, there are many paths to ecological living.

The developing world of poor countries will have converged on the same approximate level of resource use and residual production as the rich, and will have ceased to exist as a separate category of humanity. This is a level that is more than adequate for our material needs, and hence one at which infant mortality rates and birth rates are universally low and in balance. We are a species that has been given an enormous amount of power; what remains for us to do is to learn to use it in harmony with the rest of the biosphere while we still have the opportunity to do so. Clearly we are a very long way from this level of achievement today. Nor are we moving in the right direction.

12.7 Conclusion

There was a more optimistic time – not so long ago – when environmentalists exhorted their fellow citizens to take up the challenge of becoming 'stewards of the earth'. As James Lovelock remarked, he 'would sooner expect a goat to succeed as a gardener than expect humans to become responsible stewards of the Earth' (Lovelock 1991: 186). Any honest appraisal would find our credentials for the steward's job to be rather spare. Indeed, historians of the future will find it hard to understand how we created this mess, and having created it how slow we were to put it right.

Any such historian would have to understand a number of peculiarities of modern Western society, including the following:

- Politicians usually want to hold on to power.
- Scientists want to protect their reputation for cautious objectivity (and hence their research funding).
- Business people want to protect the bottom line, the share price, and dividends on a quarterly basis, as well as their bonus on an annual basis.
- Citizens want guarantees for their financial security, health and safety at no extra cost to themselves.

Those are the constraints in the rich democracies where the technical means are available to chart a completely different, ecological path at little, or even no, extra cost. Conditions are considerably more difficult in poorer countries where survival for many is still a daily struggle. Change will not come until these conflicting goals and this uneven distribution of resources begin to move towards some point of reconciliation.

In the meantime, does the motorist sitting in a traffic jam on the M25 or the Periphérique somehow think that he or she can step out of the natural cycle of life? Is the Western world so immersed in the triumph of image over substance that the virtual world appears to have become the real world?

Climate change will make a difference to the way human society functions. We must hope that it will not lead to a death toll on the scale of the Black Death. Climate change will also cause a shift in the locus of power – just as the Black Death caused a labour shortage in fourteenth century Europe, which may in turn have led to a decline in the feudal system that had kept labour tied to the land. Although we cannot predict the shape of the future it seems reasonable to assume that changes will be significant and hence people and institutions that prepare for climate change will fare better than those that do not.

At this point we must return to the opening observation in this book concerning the divorce of Western urban society from biological reality. It is very difficult for people to accept that the climate may begin to change in an unpredictable fashion if they are largely unaware of how the climate functions in the first place. In Britain it is still common to refer to warm, dry weather as 'good weather' and cool, damp days as 'bad'. Will a hot, arid summer be even better? Where does water come from if not from those cool, damp days?

Many people understand all these issues and know that our demands are dangerously out of kilter with what the world can supply on a sustainable basis. Some of those people even live in cities and know that urban systems can, and must, be redesigned to function in keeping with the ecosystems on which they depend. Once this knowledge is linked from city to city, people may begin to see opportunities for change, where before they saw only constraints. For virtually every environmentally sound goal that we might pursue there are multiple benefits to be gained. Once this is widely understood, then the ecological city will become a reality.

12.8 Websites

1. ICLEI for the Aalborg Charter text:
   http://www.iclei.org/europe/echarter.htm

2. UK Department of the Environment, Transport and the Regions, Local Agenda 21:
   http://www.environment.detr.gov.uk/sustainable/la21/policy/sec3.htm
   (from which Table 12.1 is taken)

3. United Nations Environment Programme, Geneva: *http://www.unep.ch/*

4. World Bank Group: *http://www.worldbank.com/*

12.9 Further reading

Atkinson, A (1991). *Principles of Political Ecology*. London, Belhaven Press.

Daly, H E and Cobb, J B (1989). *For the Common Good*. London, Green Print.

Freire, M and Stren, R, eds (2001). *The Challenge of Urban Government: Policies and Practices*. Washington DC, The World Bank Institute, with The Centre for Urban and Community Studies, University of Toronto.

Head, I L (1991). *On a Hinge of History: The Mutual Vulnerability of South and North*. Toronto, University of Toronto Press.

O'Sullivan, E (1999). *Transformative Learning: Educational Vision for the 21st Century*. Toronto, an OISE/UT book published in association with the University of Toronto Press and Zed Books, London.

Stefanovic, I L (2000). *Safeguarding Our Common Future: Rethinking Sustainable Development*. Albany, State University of New York Press.

Vanderburg, W H (2000). *The Labyrinth of Technology*. Toronto, University of Toronto Press.

APPENDIX 1

Charter of European Cities and Towns Towards Sustainability (The Aalborg Charter)

(as approved by the participants at the European Sustainable Cities and Towns Campaign in Aalborg, Denmark on 27 May 1994)

Part 1 Consensus Declaration: European Cities & Towns Towards Sustainability

1.1 The Role of European Cities and Towns

We, European cities & towns, signatories of this Charter, state that in the course of history, our towns have existed within and outlasted empires, nation states, and regimes and have survived as centres of social life, carriers of our economies, and guardians of culture, heritage and tradition. Along with families and neighbourhoods, towns have been the basic elements of our societies and states. Towns have been the centres of industry, craft, trade, education and government.

We understand that our present urban lifestyle, in particular our patterns of division of labour and functions, land-use, transport, industrial production, agriculture, consumption, and leisure activities, and hence our standard of living, make us essentially responsible for many environmental problems humankind is facing. This is particularly relevant as 80 percent of Europe's population live in urban areas.

We have learnt that present levels of resource consumption in the industrialised countries cannot be achieved by all people currently living, much less by future generations, without destroying the natural capital.

We are convinced that sustainable human life on this globe cannot be achieved without sustainable local communities. Local government is close to where environmental problems are perceived and closest to the citizens and shares responsibility with governments at all levels for the well-being of humankind and nature. Therefore, cities and towns are key players in the process of changing lifestyles, production, consumption and spatial patterns.

1.2 The Notion and Principles of Sustainability

We, cities & towns, understand that the idea of sustainable development helps us to base our standard of living on the carrying capacity of nature. We seek to achieve social justice, sustainable economies, and environmental sustainability. Social justice will necessarily have to be based on economic sustainability and equity, which require environmental sustainability.

Environmental sustainability means maintaining the natural capital. It demands from us that the rate at which we consume renewable material, water and energy resources does not exceed the rate at which the natural systems can replenish them, and that the rate at which we consume non-renewable resources does not exceed the rate at which sustainable renewable resources are replaced. Environmental sustainability also means that the rate of emitted pollutants does not exceed the capacity of the air, water, and soil to absorb and process them.

Furthermore, environmental sustainability entails the maintenance of biodiversity; human health; as well as air, water, and soil qualities at standards sufficient to sustain human life and wellbeing, as well as animal and plant life, for all time.

1.3 Local Strategies Towards Sustainability

We are convinced that the city or town is both the largest unit capable of initially addressing the many urban architectural, social, economic, political, natural resource and environmental imbalances damaging our modern world and the smallest scale at which problems can be meaningfully resolved in an integrated, holistic and sustainable fashion. As each city is different, we have to find our individual ways towards sustainability. We shall integrate the principles of sustainability in all our policies and make the respective strengths of our cities and towns the basis of locally appropriate strategies.

1.4 Sustainability as a Creative, Local, Balance-Seeking Process

We, cities & towns, recognise that sustainability is neither a vision nor an unchanging state, but a creative, local, balance-seeking process extending into all areas of local decision-making. It provides ongoing feedback in the management of the town or city on which activities are driving the urban ecosystem towards balance and which are driving it away. By building the management of a city around the information collected through such a process, the city is understood to work as an organic whole and the effects of all significant activities are made manifest. Through such a process the city and its citizens may make informed choices. Through a management process rooted in sustainability, decisions may be made which not only represent the interests of current stakeholders, but also of future generations.

1.5 Resolving Problems by Negotiating Outwards

We, cities & towns, recognise that a town or city cannot permit itself to export problems into the larger environment or to the future. Therefore, any problems or imbalances within the city are either brought towards balance at their own level or absorbed by some larger entity at the regional or national level. This is the principle of resolving problems by negotiating outwards. The implementation of this principle will give each city or town great freedom to define the nature of its activities.

1.6 Urban Economy Towards Sustainability

We, cities & towns, understand that the limiting factor for economic development of our cities and towns has become natural capital, such as atmosphere, soil, water and forests. We must therefore invest in this capital. In order of priority this requires

1. investments in conserving the remaining natural capital, such as groundwater stocks, soil, habitats for rare species;
2. encouraging the growth of natural capital by reducing our level of current exploitation, such as of non-renewable energy;
3. investments to relieve pressure on natural capital stocks by expanding cultivated natural capital, such as parks for inner-city recreation to relieve pressure on natural forests; and
4. increasing the end-use efficiency of products, such as energy-efficient buildings, environmentally friendly urban transport.

1.7 Social Equity for Urban Sustainability

We, cities & towns, are aware that the poor are worst affected by environmental problems (such as noise and air pollution from traffic, lack of amenities, unhealthy housing, lack of open space) and are least able to solve them. Inequitable distribution of wealth both causes unsustainable behaviour and makes it harder to change. We intend to integrate people's basic social needs as well as healthcare, employment and housing programmes with environmental protection. We wish to learn from initial experiences of sustainable lifestyles, so that we can work towards improving the quality of citizens' lifestyles rather than simply maximising consumption.

We will try to create jobs which contribute to the sustainability of the community and thereby reduce unemployment. When seeking to attract or create jobs we will assess the effects of any business opportunity in terms of sustainability in order to encourage the creation of long-term jobs and long-life products in accordance with the principles of sustainability.

1.8 Sustainable Land-Use Patterns

We, cities & towns, recognise the importance of effective land-use and development planning policies by our local authorities which embrace the strategic environmental assessment of all plans. We should take advantage of the scope for providing efficient public transport and energy which higher densities offer, while maintaining the human scale of development. In both undertaking urban renewal programmes in inner urban areas and in planning new suburbs we seek a mix of functions so as to reduce the need for mobility. Notions of equitable regional interdependency should enable us to balance the flows between city and countryside and prevent cities from merely exploiting the resources of surrounding areas.

1.9 Sustainable Urban Mobility Patterns

We, cities & towns, shall strive to improve accessibility and sustain social welfare and urban lifestyles with less transport. We know that it is imperative for a sustainable city to reduce enforced mobility and stop promoting and supporting the unnecessary use of motorised vehicles. We shall give priority to ecologically sound means of transport (in particular walking, cycling, public transport) and make a combination of these means the centre of our planning efforts. Motorised individual means of urban transport ought to have the subsidiary function of facilitating access to local services and maintaining the economic activity of the city.

1.10 Responsibility for the Global Climate

We, cities & towns, understand that the significant risks posed by global warming to the natural and built environments and to future human generations require a response sufficient to stabilise and then to reduce emissions of greenhouse gases into the atmosphere as soon as possible. It is equally important to protect global biomass resources, such as forests and phytoplankton, which play an essential role in the earth's carbon cycle. The abatement of fossil fuel emissions will require policies and initiatives based on a thorough understanding of the alternatives and of the urban environment as an energy system. The only sustainable alternatives are renewable energy sources.

1.11 Prevention of Ecosystems Toxification

We, cities & towns, are aware that more and more toxic and harmful substances are released into the air, water, soil, food, and are thereby becoming a growing threat to human health and the ecosystems. We will undertake every effort to see that further pollution is stopped and prevented at source.

1.12 Local Self-Governance as a Pre-Condition

We, cities & towns, are confident that we have the strength, the knowledge and the creative potential to develop sustainable ways of living and to design and manage our cities towards sustainability. As democratically elected representatives of our local communities we are ready to take responsibility for the task of reorganising our cities and towns for sustainability. The extent to which cities and towns are able to rise to this challenge depends upon their being given rights to local self-governance, according to the principle of subsidiarity. It is essential that sufficient powers are left at the local level and that local authorities are given a solid financial base.

1.13 Citizens as Key Actors and the Involvement of the Community

We, cities & towns, pledge to meet the mandate given by Agenda 21, the key document approved at the Earth Summit in Rio de Janeiro, to work with all sectors of our communities – citizens, businesses, interest groups – when developing our Local Agenda 21 plans. We recognise the call in the European Union's Fifth Environmental Action Programme 'Towards Sustainability' for the responsibility for the implementation of the programme to be shared among all sectors of the community. Therefore, we will base our work on cooperation between all actors involved. We shall ensure that all citizens and interested groups have access to information and are able to participate in local decision-making processes. We will seek opportunities for education and training for sustainability, not only for the general population, but for both elected representatives and officials in local government.

1.14 Instruments and Tools for Urban Management
Towards Sustainability

We, cities & towns, pledge to use the political and technical instruments and tools available for an ecosystem approach to urban management. We shall take advantage of a wide range of instruments including those for collecting and processing environmental data; environmental planning; regulatory, economic, and communication instruments such as directives, taxes and fees; and mechanisms for awareness raising including public participation. We seek to establish new environmental budgeting systems which allow for the management of our natural resources as economically as our artificial resource, 'money'.

We know that we must base our policy-making and controlling efforts, in particular our environmental monitoring, auditing, impact assessment, accounting, balancing and reporting systems, on different types of indicators, including those of urban environmental quality, urban flows, urban patterns, and, most importantly, indicators of an urban systems sustainability.

We, cities & towns, recognise that a whole range of policies and activities yielding positive ecological consequences have already been successfully applied in many cities through Europe. However, while these instruments are valuable tools for reducing the pace and pressure of unsustainability, they do not in and of themselves reverse society's unsustainable direction. Still, with this strong existing ecological base, the cities are in an excellent position to take the threshold step of integrating these policies and activities into the governance process for managing local urban economies through a comprehensive sustainability process. In this process we are called on to develop our own strategies, try them out in practice and share our experiences.

Part 2 The European Sustainable Cities and Towns Campaign

We, European cities & towns, signatories of this charter, shall move forward together towards sustainability in a process of learning from experience and successful local examples. We shall encourage each other to establish long-term local action plans (Local Agendas 21), thereby strengthening inter-authority cooperation, and relating this process to the European Union's actions in the field of the urban environment.

We hereby initiate the European Sustainable Cities and Towns Campaign to encourage and support cities and towns in working towards sustainability. The initial phase of this Campaign shall be for a two-year period, after which progress shall be assessed at a Second European Conference on Sustainable Cities & Towns to be held in 1996.

We invite every local authority, whether city, town or county and any European network of local authorities to join the Campaign by adopting and signing this Charter.

We request all the major local authority networks in Europe to undertake the coordination of the Campaign. A Coordinating Committee shall be established of representatives of these networks. Arrangements will be made for those local authorities which are not members of any network.

We foresee the principal activities of the Campaign to be to:

- facilitate mutual support between European cities and towns in the design, development and implementation of policies towards sustainability;
- collect and disseminate information on good examples at the local level;
- promote the principle of sustainability in other local authorities;
- recruit further signatories to the Charter;
- organise an annual 'Sustainable City Award';
- formulate policy recommendations to the European Commission;

- provide input to the Sustainable Cities Reports of the Urban Environment Expert Group;
- support local policy-makers in implementing appropriate recommendations and legislation from the European Union;
- edit a Campaign newsletter.

These activities will require the establishment of a Campaign Coordination Committee. We shall invite other organisations to actively support the Campaign.

Part 3 Engaging in the Local Agenda 21 processes: Local Action Plans Towards Sustainability

We, European cities & towns, signatories of this Charter, pledge by signing this Charter and joining the European Sustainable Cities and Towns Campaign that we will seek to achieve a consensus within our communities on a Local Agenda 21 by the end of 1996. This will meet the mandate established by Chapter 28 of Agenda 21 as agreed at the Earth Summit in Rio in June 1992. By means of our individual local action plans we shall contribute to the implementation of the European Union's Fifth Environmental Action Programme 'Towards Sustainability'. The Local Agenda 21 processes shall be developed on the basis of Part One of this Charter.

We propose that the process of preparing a local action plan should include the following stages:

- recognition of the existing planning and financial frameworks as well as other plans and programmes;
- the systematic identification, by means of extensive public consultation, of problems and their causes;
- the prioritisation of tasks to address identified problems;
- the creation of a vision for a sustainable community through a participatory process involving all sectors of the community;
- the consideration and assessment of alternative strategic options;
- the establishment of a long-term local action plan towards sustainability which includes measurable targets;
- the programming of the implementation of the plan including the preparation of a timetable and statement of allocation of responsibilities among the partners;
- the establishment of systems and procedures for monitoring and reporting on the implementation of the plan.

We will need to review whether the internal arrangements of our local authorities are appropriate and efficient to allow the development of the Local Agenda 21 processes, including long-term local action plans towards sustainability. Efforts may be needed to improve the capacity of the organisation which will include reviewing the political arrangements, administrative procedures, corporate and inter-disciplinary working, human resources available and inter-authority cooperation including associations and networks.

Signed in Aalborg, Denmark, 27 May 1994
More than 120 European cities, towns and counties have up to now signed the Aalborg Charter and thereby joined the European Sustainable Cities and Towns Campaign.

Available from: *http://www.iclei.org/europe/echarter.htm*

APPENDIX 2

Final Nagoya Declaration

PREAMBLE

WE, THE REPRESENTATIVES of 145 local government organisations from 29 nations around the world,

PARTICIPATING in the Cities for Climate Protection (CCP) World Summit, the 4th Local Government Leaders' Summit on Climate Change, sponsored by the City of Nagoya, Aichi Prefecture, Japan, and the International Council for Local Environmental Initiatives (ICLEI) and held in Nagoya, Japan, November 26–28, 1997;

AFFIRMING the important role that the CCP – a global campaign whose members include 201 local governments worldwide representing approximately 100 million people and accounting for almost 5% of global carbon dioxide emissions – is playing in assisting national governments to implement the UN Framework Convention on Climate Change (FCCC);

REFLECTING the conclusion of the IPCC Assessment Report that stabilisation of the global climate may require reductions in greenhouse gas emissions by more than 50%;

SHARING grave concern about the threat of climate change to cities and to life on the planet, and determined to undertake initiatives to reduce greenhouse gas emissions with the on-going support of ICLEI, who represents our common voice in the United Nations and the Conference of the Parties (COP) to the FCCC and its Subsidiary Bodies,

DO HEREBY PRESENT this Declaration to the Third Meeting of the COP to the FCCC taking place in Kyoto, Japan, December 1–10, 1997.

1.0 A Global Effort By Local Governments

1.1 Cities and urban areas are especially at risk from the potential effects, both direct and indirect, of climate change. Coastal cities are threatened by sea level rise, as well as more intense storms, especially in extra-tropical regions. Public health is also at great risk. As seasonal temperatures rise, especially in the summer, many residents of our cities will be threatened by heat-related illness, respiratory disease, and various infectious diseases caused by the spread of insect and rodent vectors. Ground-level ozone pollution is a serious problem in many of our cities, and its formation is also sensitive to elevated temperature.

1.2 Local governments in most parts of the world are responsible for land use, waste management, trees and parks, transportation infrastructure, building and construction codes, energy utilities, and public education. They can employ these powers to significantly reduce energy use and greenhouse gas emissions.

1.3 Local governments, in response to the threat of climate change, have taken early action to reduce local emissions of greenhouse gases. Among the participants in the Cities for Climate Protection campaign, 39 local governments have committed themselves to carbon dioxide reductions in the range of 15%–30%.

1.4 Among participants in the Cities for Climate Protection, 34 local governments have completed all five milestones required, including implementation of a Local Action Plan that aims to reduce greenhouse gas emissions. These milestones are:

- an energy and emissions baseline inventory for municipal operations and the wider community,
- estimation of an energy and emissions forecast for the target year 2010 or 2015,
- establishment of a greenhouse gas emissions reduction target,
- development of and obtaining local council approval for the Local Action Plan,
- implementation of policies and measures.

Collectively, 62 local governments reported to ICLEI in a recent survey that they have reduced their cumulative emissions by 42 million tonnes during the period 1990–1996.

1.5 Many local governments that have committed to reducing their emissions by at least 20% by the year 2005 or 2010 are now on trajectories to achieve their targets by investing in energy efficiency and transportation projects that reduce local energy use, as well as waste management policies that reduce methane emissions.

1.6 In cooperation with the private sector and NGOs, local governments are aggressively implementing strategies to reduce greenhouse gas emissions. They are enjoying multiple benefits from their initiatives that outweigh and often offset the financial costs of such measures. Such benefits include:

- lower costs of municipal operations,
- local job creation,
- better air quality and improvements in public health,
- reductions in traffic congestion,
- better urban liveability.

1.7 Inspired by the multiple benefits that local governments in developed nations are enjoying as they successfully implement greenhouse gas emissions reduction strategies, local governments in developing nations have also begun to initiate actions to reduce greenhouse gas emissions, even though their national governments are not obligated under the Framework Convention on Climate Change to reduce such emissions. Provided adequate financial resources and appropriate technologies are available, developing country cities are desirous of pursuing sustainable development strategies that reduce greenhouse gas emissions. Local governments in developing nations believe that when strategies to reduce greenhouse gas emissions address social problems and improve the standard of living locally, they are indeed worth pursuing.

2.0 Communiqué to COP 3

WE, the participants in the CCP, as well as other ICLEI members, are convinced from our concrete experiences that:

- climate change is the most serious long-term environmental threat to cities and their residents, as well as to global security;
- cities can meet significant greenhouse gas reduction targets;
- cities derive multiple benefits from greenhouse gas reductions that frequently exceed the financial costs;
- legally binding national commitments to reduce greenhouse gas emissions, if the agreed upon targets and timetables are ambitious, will significantly enhance and amplify local initiatives. On the other hand, weak national commitments risk undermining local government initiatives;
- local governments in developing nations are enthusiastic about contributing to climate protection strategies, given the multiple benefits to be enjoyed, provided adequate finances and technical assistance are made available.

THEREFORE, WE call on the Parties to the UNFCCC to:

2.1 Adopt a protocol that specifies legally binding targets and timetables for reducing greenhouse gas emissions.

2.2 Set an initial reduction target for Annex 1 countries, relative to 1990 levels, for the year 2005, with the ultimate target of 20% for the year 2010. Only such an ambitious and early target will demonstrate that national governments are making a serious effort to combat climate change and its impacts.

2.3 Urge national governments, through direct consultation with national and local organisations and municipalities, to establish processes and economic instruments which will enable the development and implementation of Local Action Plans to meet greenhouse gas reduction targets.

2.4 Persuade international agencies and development banks to grant developing country cities access to financial resources directly from the financial mechanisms associated with the UNFCCC and aid agencies such as the World Bank and regional development banks.

2.5 Recognise the benefit of a partnership approach between all spheres of government, the private sector and NGOs in undertaking climate protection measures.

2.6 Draw on the experience and expertise of local governments in implementing greenhouse gas reduction strategies by inviting local government officials to participate, through their international associations, ICLEI and IULA, on a non-voting basis in all meetings of the Conference of the Parties and its Subsidiary Bodies, alongside representatives of other international governmental organisations.

3.0 Local Government Pledge

WE, the representatives of local governments worldwide, are convinced that local governments are an effective vehicle to help achieve national greenhouse gas reduction goals and that close partnerships with our national governments will enhance our collective efforts to reduce greenhouse gas emissions;

THEREFORE, WE pledge to make maximum efforts to reduce greenhouse gas emissions at the local level, to embrace the goals of global sustainability and Local Agenda 21, and to extend our full support to our national governments and to ICLEI's CCP campaign, which will seek to:

3.1 Promote the significant reduction of greenhouse gas emissions from municipal and other public operations, including buildings, facilities, landfills, waste treatment, and water pumping stations through the use of renewables, energy efficiency, co-generation, district energy and recycling.

3.2 Promote the significant reduction of greenhouse gas emissions from community-wide activities, including transportation, housing, and commerce locally.

3.3 Promote the significant expansion of the supply and consumption of renewable energy worldwide at the municipal level.

3.4 Promote local educational initiatives and organisations to enhance public understanding of climate change, thereby improving acceptance of national government measures to reduce greenhouse gas emissions.

3.5 Recruit by the year 2000 local governments to the CCP campaign that together with existing CCP participants account for 10% of the world's carbon dioxide emissions.

3.6 Significantly expand the CCP campaign in developing countries.

28 November 1997
Available from: http://www.iclei.org/co2/nagoya.htm

APPENDIX 3

The 'Annex 1' Countries: Signatories to the Kyoto Protocol

Australia
Austria
Belgium
Bulgaria*
Canada
Croatia*
Czech Republic*
Denmark
Estonia*
European Community
Finland
France
Germany
Greece
Hungary*
Iceland
Ireland
Italy
Japan
Latvia*

Liechtenstein
Lithuania*
Luxembourg
Monaco
Netherlands
New Zealand
Norway
Poland*
Portugal
Romania*
Russian Federation*
Slovakia*
Slovenia*
Spain
Sweden
Switzerland
Ukraine*
United Kingdom of Great Britain and
    Northern Ireland
United States of America

Note: * signifies a party 'undergoing the process of transition to a market economy to which a certain degree of flexibility' will be allowed 'in the implementation of their commitments' (Articles 3.5 and 3.6 of the Kyoto Protocol)

Source: Grubb *et al.* 1999: 284, 301.

Bibliography

Adur District Council Environmental Services (1999). *Annual Report 1998/99: Adur Quality of Life*. Adur District Council, UK.

Akbari, H and Taha, H (1991). 'The impact of trees and white surfaces on residential heating and cooling energy use in four Canadian cities', *Energy, the International Journal*, 17, 2: 141–9.

Akbari, H, Davis, S, Dorsano, S, Huang, J and Winnett, S, eds (1992). *Cooling Our Communities: A Guidebook on Tree Planting and Light-Colored Surfacing*. Washington DC, United States Environmental Protection Agency.

Akbari, H, Konopacki, S and Pomerantz, M (1999). 'Cooling energy savings potential of reflective roofs for residential and commercial buildings in the United States', *Energy*, 24: 391–407.

Aldred, C (1995). 'Rules may increase UK asbestos claims: more mesothelioma victims given grounds to make claims', *Business Insurance*, 27 November: 31.

Ashton, J, ed (1992). *Healthy Cities*. Milton Keynes, Open University Press.

Association of British Insurers (2000). *Subsidence: A Global Perspective*. General Insurance: Research Report No 1. London, ABI.

Atkinson, A (1991). *Principles of Political Ecology*. London, Belhaven Press.

Bakker, K (2000). 'Privatising water, producing scarcity: the Yorkshire Drought of 1995', *Economic Geography*, 76, 1: 4–27.

Barton, H, ed (2000). *Sustainable Communities: The Potential for Eco-Neighbourhoods*. London, Earthscan.

Boardman, B (1991). *Fuel Poverty: From Cold Homes to Affordable Warmth*. London, Belhaven Press.

Boardman, B, Fawcett, T, Griffin, H, Hinnells, M, Lane, K and Palmer, J (1997). *DECADE – Domestic Equipment and Carbon Dioxide Emissions. 2 MtC – Two Million Tonnes of Carbon*. Oxford, University of Oxford, Environmental Change Unit.

Borkiewicz, J, Mieczkowska, E, Aleksandrowicz, A and Leitman, J (1991). *Environmental Profile of Katowice*. Washington DC, The World Bank.

Bower, B T, ed (1977). *Regional Residuals Environmental Quality Management Modeling*. Washington DC, Resources for the Future.

Boyden, S, Millar, S, Newcombe, K and O'Neill, B (1981). *The Ecology of a City and its People: The Case of Hong Kong.* Canberra, Australian National University Press.

Boyle, G, ed (1996). *Renewable Energy: Power for a Sustainable Future.* Milton Keynes, Oxford University Press in association with the Open University.

Brook, J (1999). 'Health effects of atmospheric change in the Toronto–Niagara Region', in *Atmospheric Change in the Toronto–Niagara Region: Towards an Integrated Understanding of Science, Impacts and Responses.* Eds B Mills and L Craig. Toronto, Environment Canada/Ontario Ministry of the Environment/University of Toronto: 81–7.

Brower, M (1993). *Cool Energy: Renewable Solutions to Environmental Problems.* Cambridge, MIT Press.

Browne, J (1997). 'Climate change: the new agenda.' A presentation to Stamford University, California, May 1997.

Caccia, C (1996). 'Climate change, CO_2 and subsidies: change the tax system', Appendix IV in *Proceedings of the Climate Variability, Atmospheric Change and Human Health Conference,* Toronto, 4–5 November. Eds T Hancock, P Morrison and S Peck. Toronto: Pollution Probe, York University, University of Toronto and Environment Canada.

Calthrop, E J (1995). 'The external costs of road transport fuel: should the fiscal stance towards diesel be altered?' Unpublished MSc thesis, University College, London.

Canadian Urban Institute (1993). *Cooling Buildings in Downtown Toronto.* Final Report of the Deep Lake Water Cooling Investigation Group. Toronto, Canadian Urban Institute.

Canzi, M (1998). 'A role for the unplugged coach-house in residential intensification.' Unpublished paper, University of Toronto, Department of Geography and Program in Planning.

Case, P (1999). *Environmental Risk Management and Corporate Lending: A Global Perspective.* Cambridge, Woodhead Publishing.

CHEJ (Center for Health, Environment and Justice) (2000). *A History of the Love Canal.* Available at *www.chej.org/lcindex.html*

Chia, M (1997). 'Financial implications of lead toxicity: a critical assessment.' Unpublished paper, Institute for Environmental Studies, University of Toronto.

City of Toronto (1991). *The Changing Atmosphere: Strategies for Reducing Carbon Dioxide Emissions. Volume 1, Policy Overview; Volume 2, Technical Volume.* Toronto, Special Advisory Committee on the Environment.

City of Toronto (1998). *Smog: Make It or Break It – a Citizens' Action Guide with Ideas, Inspirations and Success Stories.* Toronto, City of Toronto.

Connell, D, Lam, P, Richardson, B and Wu, R (1999). *Introduction to Ecotoxicology.* Oxford, Blackwell Science.

Daly, H E and Cobb, J B (1989). *For the Common Good.* London, Green Print.

Davis, M L and Cornwell, D A (1998). *Introduction to Environmental Engineering* (third edition). Boston, McGraw-Hill.

De Sousa, C (2000). 'The brownfield problem in Canada: issues, approaches and solutions.' Unpublished PhD thesis, University of Toronto, Department of Geography.

Department of the Environment, Transport and the Regions (1998). *Cryptosporidium in Water Supplies.* London, Report to the Secretaries of State for DETR, and Health. Available at *www.dwi.detr.gov.uk/pubs/bouchier/bou00.htm*

Department of the Environment, Transport and the Regions (2000). *Asbestos and Man-made Mineral Fibres in Buildings: Practical Guidance.* Available at *www.environment.detr.gov.uk/asbestos/guide/index.htm*

Dougan, T (1998). *Sustainable Housing: Some Lessons from Milton Keynes.* Milton Keynes, Network for Alternative Technology and Technology Assessment.

Douglas, I (1983). *The Urban Environment.* London, Edward Arnold.

Drinking Water Inspectorate, UK (2000). *Summary Report for 1998 on Thames Water Utilities Ltd.* Available at *http://www.dwi.detr.gov.uk/pubs/coreport/tha98.htm*

Duffy, N (1999). 'The relationship between urban design and the potential leaf area in the urban forest. MScF thesis, University of Toronto, Faculty of Forestry.

Economist, The (1992a). 'Let them eat pollution', 8 February: 82.

Economist, The (1992b). 'Summers on sustainable growth', 30 May: 65.

Elliott, D (1997). *Energy, Society and Environment: Technology for a Sustainable Future.* London, Routledge.

Environment Canada (1996). *Climate Variability, Atmospheric Change and Human Health Conference Proceedings.* Downsview, Ontario, Environment Canada.

Etkin, D (1999). 'Risk transference and related trends: driving forces towards more mega-disasters', *Environmental Hazards*, 1: 69–75.

Fawcett, T, Lane, K and Boardman, B with other contributors (2000). *Lower Carbon Futures for European Households.* Oxford, University of Oxford, Environmental Change Institute.

Ferguson, B K (1998). *Introduction to Stormwater: Concept, Purpose and Design.* New York, Wiley.

Firor, J (1990). *The Changing Atmosphere: The Global Challenge.* New Haven, Yale University Press.

Freire, M and Stren, R, eds (2001). *The Challenge of Urban Government: Policies and Practices.* Washington DC, The World Bank Institute, with The Centre for Urban and Community Studies, University of Toronto.

Gaia Trust and Findhorn Foundation (1996). *Eco-Villages and Sustainable Communities: Models for 21st Century Living.* Findhorn, Findhorn Press.

Gilbert, R, Stevenson, D, Girardet, H and Stren, R E (1996). *Making Cities Work: The Role of Local Authorities in the Urban Environment.* London, Earthscan.

Girardet, H (1992). *The Gaia Atlas of Cities: New Directions for Sustainable Urban Living.* New York, Anchor Books published by Doubleday.

Girardet, H (1999). *Creating Sustainable Cities.* Totnes, Green Books for the Schumacher Society. Schumacher Briefing No 2.

Goodwin, P (2000). 'Current thinking on traffic control. Part I: The theoretical background', in *Transport and the Future of Oxford.* Ed E Townsend-Coles. Oxford, Oxford Civic Society: 14–18.

Goubert, J-P (1989). *The Conquest of Water: The Advent of Health in the Industrial Age.* Cambridge, Polity Press.

Goudie, A (1986). *The Human Impact on the Natural Environment.* Cambridge, MIT Press.

Gould, S J (1999). *Leonardo's Mountain of Clams and the Diet of Worms.* London, Vintage.

Graedel, T E (1999). 'Environmentally superior buildings from birth to death', in *Reshaping the Built Environment.* Ed C J Kibert. Washington DC, Island Press: 259–75.

Gray, M J (1999). 'Assessment of water supply and associated matters in relation to the incidence of cryptosporidiosis in West Herts and north London in February and March 1997.' Report to the Drinking Water Inspectorate. Available at *www.dwi.detr.gov.uk/pubs/sumrep/clane3a.htm*

Grubb, M (1995). *Renewable Energy Strategies for Europe. Volume 1 – Foundations and Context.* London, Royal Institute of International Affairs and Earthscan Publications.

Grubb, M with Vrolijk, C and Brack, D (1999). *The Kyoto Protocol: A Guide and Assessment*. London, Royal Institute of International Affairs.

Guerra, L M (1991). 'Urban air quality and health.' Paper presented at a conference on Cities and Global Change, Toronto, 12–14 June.

Hardoy, J and Satterthwaite, D (1989). *Squatter Citizen: Life in the Urban Third World*. London, Earthscan Publications.

Hardoy, J, Mitlin, D and Satterthwaite, D (1992). *Environmental Problems in Third World Cities*. London, Earthscan.

Harvey, L D D (2000). *Climate and Global Environmental Change*. London, Prentice Hall.

Haughton, G and Hunter, C (1994). *Sustainable Cities*. Regional Policy and Development Series 7. London, Jessica Kingsley Publishers and the Regional Studies Association.

Head, I L (1991). *On a Hinge of History: The Mutual Vulnerability of South and North*. Toronto, University of Toronto Press.

Henry, J G and Heinke, G W (1989). *Environmental Science and Engineering*. Englewood Cliffs, Prentice Hall.

Higuchi, K, Yuen, C W and Shabbar, A (2000). 'Ice storm '98 in southcentral Canada and northeastern United States: a climatological perspective', *Theoretical and Applied Climatology*, 66: 61–79.

Hough, M (1989). *City Form and Natural Process*. London, Routledge.

Hough, M (1995). *Cities and Natural Process*. London, Routledge.

Houghton, J (1994). *Global Warming: The Complete Briefing*. Oxford, Lion Publishing.

Houghton, J T, Jenkins, G J and Ephraums, J J, eds (1990). *Climate Change: The IPCC Scientific Assessment*. Cambridge, Cambridge University Press.

Houghton, J T, Callander, B A and Varney, S K, eds (1992). *Climate Change 1992: The Supplementary Report to the IPCC Scientific Assessment*. Cambridge, Cambridge University Press.

Houghton, J T, Meiro Filho, L G, Callander, B A, Harris, N, Kattenberg, A and Maskell, K, eds (1996). *Climate Change 1995: The Science of Climate Change*. Cambridge, Cambridge University Press.

Howard, E (1986 (1898)). *Garden Cities of Tomorrow*. Builth Wells, Attic Books.

Hugenschmidt, H, Janssen, J, Kermode, Y and Schumacher, I (1999). 'Sustainable banking at UBS', *Greener International Management*, 27: 38–48.

Hulme, M and Jenkins, G J (1998). *Climate Change Scenarios for the United Kingdom: Scientific Report*. UK Climate Impacts Programme, Technical Report No 1. Norwich, Climate Research Unit.

Husslage, W J G (1997). 'Green/blue lungs and veins in the heart of Toronto: progress in adopting ecological approaches for the regeneration of Metro's downtown.' Unpublished paper, University of Toronto, Department of Geography and Program in Planning.

Institute for Environmental Studies and Pollution Probe (1998). *Emissions from Coal-Fired Electric Stations: Environmental Health Effects and Reduction Options*. Toronto, Institute for Environmental Studies, University of Toronto.

Institute for Environmental Studies and Environment Canada (1999). *Air Issues and Urban Ecosystems: Towards a Conceptual Model*. Report on a workshop, 22 March 1999, Toronto. Institute for Environmental Studies, University of Toronto, and Environment Canada – Ontario Region.

Insurance Bureau of Canada (1994). *Improving the Climate for Insuring Environmental Risks*. Report of the Environmental Liability Committee, August 1994. Toronto, IBC.

Insurance Bureau of Canada (1995). *Improving the Climate for Insuring Environ-*

mental Risks. Proceedings from an industry symposium, June 1995. Toronto, IBC.

Insurance Services Office (1997). *The Wildland/Urban Fire Hazard*. New York, ISO.

International Council for Local Environmental Initiatives (2000). *Cities at Risk: Assessing the Vulnerability of the United States Cities to Climate Change*. Cities for Climate Protection Program. Available from *http://www.iclei.org/co2/cartextonly.htm*

Jackson, A R W and Jackson, J M (1996). *Environmental Science: The Natural Environment and Human Impact*. London, Longman.

Jackson, T (1992). 'Renewable energy: summary paper for the renewable series', *Energy Policy*, 20, 9: 861–83.

Jacobs, J (1961). *The Death and Life of Great American Cities*. New York.

Jankovic, M S and Amott, N (1998). 'Greening parking lots in the City of Toronto.' Unpublished paper. Institute for Environmental Studies, University of Toronto.

Jenks, M, Burton, E and Williams, K, eds (1996). *The Compact City: A Sustainable Urban Form*. London, E and F N Spon.

Jeucken, M H A and Bouma, J J (1999). 'The changing environment of banks', *Greener Management International*, 27, Autumn: 21–35.

Kerry, M, Kelk, G, Etkin, D, Burton, I and Kalhok, S (1998). 'Glazed over: Canada copes with the ice storm of 1998', *Environment*, 41, 1: 6–11, 28–33.

Khan, E (1998). 'Reducing transportation emissions within the Toronto–Niagara Region.' Unpublished paper. Institute for Environmental Studies, University of Toronto.

Kibert, C J (1999). *Reshaping the Built Environment: Ecology, Ethics and Economics*. Washington DC, Island Press.

Krause, E (1997). 'Ecological footprints, climate change, and sustainable development in the Greater Toronto area.' MA thesis, University of Toronto, Department of Geography and Institute for Environmental Studies.

Kumar, A (1998). 'Technical mitigation options for addressing CO_2 reduction options from the energy sector within the Toronto–Niagara region.' Unpublished paper. Institute for Environmental Studies, University of Toronto.

Kulkarni, T (2000). 'Urban infrastructure floods in southern Ontario: a methodology to determine causality' (Part One) *Assurances*, 68, 1: 1–20; (Part Two) 68, 2: 179–98.

Lecomte, E L, Pang, A W and Russell, J W (1998). *Ice Storm '98*. Toronto, Institute for Catastrophic Loss Reduction, Insurance Council of Canada.

Lecomte, P (1999). *Polluted Sites: Remediation of Soils and Groundwater*. Rotterdam, A A Balkema.

Lee, S (1998). 'Hydrological metabolism and water resource management of the Beijing metropolitan region in the Hai River basin.' Unpublished MSc thesis, University of Toronto, Department of Geography and Institute for Environmental Studies.

Leggett, J (1999). *The Carbon War: Dispatches from the End of the Oil Century*. Harmondsworth, Penguin Books.

Local Government Management Board (UK) (1994). *The Sustainability Indicators Research Project: Indicators and LA21 – A Summary*. London, Department of the Environment, Transport and the Regions.

Logan, J, Frank, A, Feng, J and Indu, J (1999). *Climate Action in the United States and China*. Advanced International Studies Unit (Battelle Memorial Institute) and the Woodrow Wilson International Center for Scholars, Washington DC.

Loughborough, K (1999). 'Perspectives from the City of Toronto public works', in *Atmospheric Change in the Toronto–Niagara Region: Towards an Integrated Understanding of Science, Impacts and Responses*. Eds B Mills and L Craig.

Toronto, Environment Canada/Ontario Ministry of the Environment/University of Toronto. Available at *http://www1.tor.ec.gc.ca/earg/tnrs/htm*

Lovelock, J (1991). *Gaia: The Practical Science of Planetary Medicine*. London, Gaia Books Ltd.

Low, N, Gleeson, B, Elander, I and Lidskog, R, eds (2000). *Consuming Cities: The Urban Environment in the Global Economy after the Rio Declaration*. London, Routledge.

Lyle, J T (1999). 'Landscape: source of life or liability', in *Reshaping the Built Environment: Ecology, Ethics and Economics*. Ed C J Kibert. Washington DC, Island Press.

McHarg, I L (1969). *Design with Nature*. Garden City, NY, Doubleday and Co.

McGranahan, G and Satterthwaite, D (2000). 'Environmental health or ecological sustainability? Reconciling the brown and green agendas', *Proceedings of the 17th Inter-Schools Conference on Sustainable Cities – The Urban Agenda in Developing Countries*. Ed R Zetter. Oxford Brookes University, 3–4 April: 27–41.

Maddison, D, Pearce, D, Johansson, O, Calthrop, E, Litman, T and Verhoef, E (1996). *The True Cost of Road Transport*. Blueprint No 5. London, Earthscan.

Mackay, D (1991). *Multimedia Environmental Models: The Fugacity Approach*. Chelsea, MI, Lewis Publishers.

Martinez-Alier, J (1990). *Ecological Economics: Energy, Environment and Society*. Oxford, Basil Blackwell.

Maxwell, B, Mayer, N and Street, R, (1997). *The Canada Country Study: Climate Impacts and Adaptation. National Summary for Policy Makers*. Ottawa, Environment Canada.

Meade, M, Florin, J and Gesler, W (1988). *Medical Geography*. New York, Guilford Press.

Meadows, D, Meadows, D and Randers, J (1992). *Beyond the Limits: Global Collapse or a Sustainable Future?* London, Earthscan.

Meyer, A and Cooper, T (2000). 'Why convergence and contraction are the key', *Environmental Finance*, May: 19–21.

Mills, B and Craig, L, eds (1999). *Atmospheric Change in the Toronto–Niagara Region: Towards an Integrated Understanding of Science, Impacts and Responses*. Toronto, Environment Canada/Ontario Ministry of the Environment/University of Toronto.

Mortsch, L and Mills, B, eds (1996). *Great Lakes–St Lawrence Basin Project on Adapting to the Impacts of Climate Change and Variability*. Progress Report No 1. Ottawa, Environment Canada, Adaptation and Impacts Research Group. Available at *http://www1.tor.ec.gc.ca/airg/pubs/execsum1.htm*

Munich Re (1999). *Annual Review of Natural Catastrophes 1998*. Munich, Munich Re.

Munich Re (1997). *Flooding and Insurance*. Munich, Munich Re.

Munn, R E, ed (1994). 'Looking ahead: the inclusion of long term futures in cumulative environmental assessment', *Environmental Monograph No 11*, Institute for Environmental Studies, University of Toronto.

Munn, R E, ed (1997). 'Atmospheric change in Canada: assessing the whole as well as the parts. Proceedings of a workshop held in Toronto, 27–28 March 1995', *Environmental Monitoring and Assessment*, 46, 1–2, June. Also available from *www.utoronto.ca/env/env_mon.htm*

Munn, R E and Maarouf, A (1997). 'Atmospheric issues in Canada', *The Science of the Total Environment*, 203: 1–11.

Munn, R E, Maarouf, A and Cartmale, L (1997). 'Atmospheric change in Canada: assessing the whole as well as the parts', *Environmental Monitoring and Assessment*, 46: 1–4.

Munn, R E and Wheaton, P, eds (1997). 'Proceedings: environmental assessment follow-up and monitoring workshop, 15–16 April 1997', *Environmental Mono-graph No 14*, Institute for Environmental Studies, University of Toronto.

Newbery, D (1995). 'Royal Commission Report on Transport and the Environment – an economic critique.' Unpublished memo. University of Cambridge.

Nicholls, M (2000). 'Ontario looks across borders', *Environmental Finance*, April: 6.

NOAA (1995). *Natural Disaster Survey Report: July 1995 Heat Wave*. Washington DC, US Department of Commerce.

NOAA/USDA Joint Agricultural Weather Facility, Tropical Prediction Center (2000). 'Climatic anomalies in 1999', *Environmental Finance*, March: 18–19.

Noorman, K J and Uiterkamp, T S (1998). *Green Households? Domestic Consumer, Environment and Sustainability*. London, Earthscan.

Odum, H T (1971). *Environment, Power and Society*. New York, Wiley Inter-science.

Odum, H T (1983). *Systems Ecology: An Introduction*. New York, Wiley.

Ogilvie, K (1999). 'Mitigation: if we thought we had to, we could', in *Atmospheric Change in the Toronto–Niagara Region: Towards an Integrated Understanding of Science, Impacts and Responses*. Eds B Mills and L Craig. Toronto, Environment Canada/Ontario Ministry of the Environment/University of Toronto: 129–34.

Onisto, L, Krause, E and Wackernagel, M (1998). *How Big is Toronto's Ecological Footprint? Using the Concept of Appropriated Carrying Capacity for Measuring Sustainability*. United Nations Department of Public Information, DPI/1907/sd.

O'Sullivan, E (1999). *Transformative Learning: Educational Vision for the 21st Century*. Toronto, an OISE/UT book published in association with the University of Toronto Press and Zed Books, London.

Owen, S (1991). *Planning Settlements Naturally*. Chichester, Packard Publishing.

Oxford University Press (1992). *A Concise Dictionary of Chemistry*. Oxford, Oxford University Press.

Palmer, J and Boardman, B (1998). *DELIGHT – Domestic Efficient Lighting*. Oxford, University of Oxford, Environmental Change Unit.

Parry, M and other contributors (1991). *The Potential Effects of Climate Change in the United Kingdom*. London, HMSO published for the Department of the Environment.

Partner Re (1997). *Floods: Causes, Effects and Risk Assessment*. Bermuda, Partner Re.

Partner Re (1998). *Hurricane Georges, September 15–29, 1998*. Bermuda, Partner Re.

Pollution Probe and the Canadian Institute of Child Health (1998). *The Air Children Breathe: The Effects on their Health*. Downsview, Ontario, Environment Canada.

Rabinovitch, J (1992). 'Curitiba: towards sustainable urban development', *Environment and Urbanization*, 4, 2: 62–73.

Rabinovitch, J and Hoehn, J (1995). 'A sustainable urban transportation system: the "surface metro" in Curitiba, Brazil', Environmental and Natural Resources and Policy Training Project, Working Paper 19. Washington DC, USAID.

Rabinovitch, J and Leitman, J (1996). 'Urban planning in Curitiba', *Scientific American*, March: 46–53.

Ravetz, J (2000). *City – Region 2020: Integrated Planning for a Sustainable Environment*. London, Earthscan.

Richardson, N (1992). 'Canada', Chapter 6, pp. 145–67, in *Sustainable Cities: Urbanization and the Environment in International Perspective*. Eds R E Stren, R R White and J B Whitney. Boulder, Westview Press.

Robinson, P (2000). 'Canadian municipal response to climate change: a framework for analyzing barriers.' Unpublished PhD thesis, University of Toronto, Department of Geography.

Romm, J J (1999). *Cool Companies: How the Best Companies Boost Profits and Productivity by Cutting Greenhouse Gas Emissions.* London, Earthscan.

Roseland, M, ed (1997). *Eco-City Dimensions: Healthy Communities, Healthy Planet.* Gabriola Island, BC, New Society Publishers.

Roseland, M (1998). *Toward Sustainable Communities: Resources for Citizens and Their Governments.* Gabriola Island, BC, New Society Publishers.

Rowland, A J and Cooper, P (1983). *Environment and Health.* London, Edward Arnold.

Sandor, R (2000). 'DaimlerChrysler, Ford change lanes', *Environmental Finance*, March: 11. (See box: Scorecard 1 – summary of greenhouse gas emissions trading activity.)

Scott, P (2000). 'The rise of reporting', *Environmental Finance*, May: 22–3.

Sewell, J (1993). *The Shape of the City: Toronto Struggles with Modern Planning.* Toronto, University of Toronto Press.

Smith, M, Whitelegg, J and Williams, N (1998). *Greening the Built Environment.* London, Earthscan.

Southam, S F, Mills, B N, Moulton, R J and Brown, D W (1999). 'The potential impact of climate change on Ontario's Grand River Basin: water supply and demand issues', *Canadian Water Resources Journal*, 24, 4: 307–30.

Stefanovic, I L (2000). *Safeguarding Our Common Future: Rethinking Sustainable Development.* Albany, State University of New York Press.

Stren, R E and White, R R, eds (1989). *African Cities in Crisis: Managing Rapid Urban Growth.* African Modernization and Development. Boulder, Westview Press.

Stren, R E, White, R R and Whitney, J B, eds (1992). *Sustainable Cities: Urbanization and the Environment in International Perspective.* Boulder, Westview Press.

Swaigen, J (1995). *Toxic Time Bombs: The Regulation of Canada's Leaking Underground Storage Tanks.* Toronto, Montgomery.

Swiss Re (1994) Global Warming: Element of Risk. Zurich, Swiss Re.

Swiss Re (1999). *Annual Report.* Zurich, Swiss Re.

Tang, M (1996). 'The relevance of the Living Machine technology to water management in Metropolitan Toronto.' *Current Issues Paper.* University of Toronto, Program in Planning.

Thornthwaite, C W and Mather, J R (1955). *The Water Balance.* Centerton, NJ, Drexel Institute of Technology. Publications in Climatology, Vol 8, No 1.

Timmerman, P and White, R R (1997). 'Megahydropolis: coastal cities in the context of global environmental change', *Global Environmental Change*, 7, 3: 205–34.

Timmerman, P, Abate, S, Amott, N, Balog, M, Beaulieu, P, Corcoran, K, Ferguson, C, Georgakopoulos, J, Gore, C, Saletto, M, Khan, E, Kulkarni, T, Kumar, A, Lee, S H, McFayden, J, Mladinic, H and Pereverzoff, S, with Byer, P, Gibson, B and Sage, R (1999). 'The six percent solution: a contribution from the Institute for Environmental Studies to the Toronto–Niagara region study', in *Atmospheric Change in the Toronto–Niagara Region: Towards an Integrated Understanding of Science, Impacts and Responses.* Eds B Mills and L Craig. Toronto, Environment Canada/Ontario Ministry of the Environment/University of Toronto: 135–49.

Todd, J (1996). 'The technological foundations of eco-villages', in *Eco-Villages and Sustainable Communities.* Findhorn, Findhorn Press.

Todd, J (1999). 'Ecological design, Living Machines, and the purification of waters', in *Reshaping the Built Environment: Ecology, Ethics and Economics.* Ed C J Kibert. Washington DC, Island Press: 131–50.

Townsend-Coles, E, ed (2000). *Transport and the Future of Oxford.* Oxford, Oxford Civic Society.

Trumbull, W C (1999). 'The Chicago Brownfields Initiative', in *Reshaping the Built*

Environment: Ecology, Ethics and Economics. Ed C J Kibert. Washington DC, Island Press.

United Kingdom, Institute of Hydrology (1995/6). 'The 1995/6 drought', *Annual Report 1995/6.* Wallingford, NSERC: 16–19.

United States Department of Energy (1994). 'Cities cut water system energy costs.' Available from *http://www.eren.doe.gov/cities-counties/watersy.html*

United States Public Interest Research Group Educational Fund (2000). *Storm Warning: Global Warming and the Rising Costs of Extreme Weather. Executive Summary.* Washington DC.

Vanderburg, W H (2000). *The Labyrinth of Technology.* Toronto, University of Toronto Press.

Van Vuuren, D P and Smeets, E M W (2000). 'Ecological footprint of Benin, Bhutan, Costa Rica and the Netherlands', *Ecological Economics*, 43, 1: 115–30.

Wackernagel, M and Rees, W E (1996). *Our Ecological Footprint: Reducing Human Impact on the Earth.* Gabriola Island, BC, New Society Publishers.

Wackernagel, M *et al.* (1999). 'National natural capital accounting with the ecological footprint concept', *Ecological Economics*, 29, 3: 375–90.

Wade, S, Hossell, J, Hough, M and Fenn, C, eds (1999). *The Impacts of Climate Change in the South East: Technical Report.* Epsom, W S Atkins Ltd.

Watson, R T, Zinyowera, M C and Moss, R H, eds (1996). *Climate Change 1995. Impacts, Adaptations and Mitigation of Climate Change: Scientific–Technical Analyses.* Cambridge, Cambridge University Press, published for the Inter-governmental Panel on Climate Change.

Watson, R T, Zinyowera, M C, Moss, R H and Dokken, D J, eds (1998). *The Regional Impacts of Climate Change: An Assessment of Vulnerability.* Cambridge, Cambridge University Press, published for the Intergovernmental Panel on Climate Change.

Weizsacker, E von, Lovins, A B and Lovins, L H (1998). *Factor Four: Doubling Wealth, Halving Resource Use.* London, Earthscan.

White, R R (1993). *North, South and the Environmental Crisis.* Toronto, University of Toronto Press.

White, R R (1994). *Urban Environmental Management: Environmental Change and Urban Design.* Chichester, Wiley.

White, R R (2000). 'African cities and climate change: the global context for sustainable development.' *Proceedings of the 17th Inter-Schools Conference on Sustainable Cities – The Urban Agenda in Developing Countries.* Ed R Zetter. Oxford Brookes University, 3–4 April: 223–31.

White, R R and Whitney, J B (1992). 'Cities and the environment: an overview', in *Sustainable Cities: Urbanization and the Environment in International Perspective.* Eds R E Stren, R R White and J B Whitney. Boulder, Westview Press: 5–52.

White, R R and Etkin, D (1997). 'Climate change, extreme events and the Canadian insurance industry', *Natural Hazards*, 16: 136–63.

Wolfensohn, J D (1996). 'Crucibles of development', *Human Settlements – Our Planet*, 8.1. Nairobi, United Nations Environment Programme.

Wolman, A (1965). 'The metabolism of cities', *Scientific American*, September: 179–88.

World Bank (1992). *World Development Report 1992: Development and the Environment.* New York, Oxford University Press.

World Bank (1993). *World Development Report 1993: Investing in Health.* New York, Oxford University Press.

World Bank (2000). *Entering the 21st Century: World Development Report 1999/2000.* New York, Oxford University Press.

Yost, P (1999). 'Construction and demolition waste: innovative assessment and management', in *Reshaping the Built Environment: Ecology, Ethics and Economics.* Ed C J Kibert. Washington DC, Island Press: 176–95.

Zetter, R (2000). 'Market enablement or sustainable development? The conflicting paradigms of urbanisation.' Paper presented at the 17th Inter-Schools Conference on Sustainable Cities – The Urban Agenda in Developing Countries. Oxford Brookes University, 3–4 April. London, Intermediate Technology Publications, in press.

Index